Geology of the Yreka Quadrangle Siskiyou County, California

U.S. Geological Survey Bulletin

Compiled by the staff of
the United States Geological Survey

with an introduction by Kerby Jackson

Introduction

It has been nearly forty years since the United States Department of Geological Survey released it's publication "Geology of the Yreka Quadrangle, Siskiyou County, California".

The author of this work, the late Preston E. Hotz, was well known in those days for his expertise and his written contributions to the mining industry. Hotz wrote an incredibly large body of work on geology and mineral resources, mostly for the United States Geological Survey, and on a relatively diverse range of areas.

It has often been said that "*gold is where you find it*", but even beginning prospectors understand that their chances for finding something of value in the earth or in the streams of the Golden West are dramatically increased by going back to those places where gold and other minerals were once mined by our forerunners. Despite this, much of the contemporary information on local mining history that is currently available is mostly a result of mere local folklore and persistent rumors of major strikes, the details and facts of which, have long been distorted. Long gone are the old timers and with them, the days of first hand knowledge of the mines of the area and how they operated. Also long gone are most of their notes, their assay reports, their mine maps and personal scrapbooks, along with most of the surveys and reports that were performed for them by private and government geologists. Even published books such as this one are often retired to the local landfill or backyard burn pile by the descendents of those old timers and disappear at an alarming rate. Despite the fact that we live in the so-called "Information Age" where information is supposedly only the push of a button on a keyboard away, true insight into mining properties remains illusive and hard to come by, even to those of us who seek out this sort of information as if our lives depend upon it. Without this type of information readily available to the average independent miner, there is little hope that our metal mining industry will ever recover.

This important volume and others like it, are being presented in their entirety again, in the hope that the average prospector will no longer stumble through the overgrown hills and the tailing strewn creeks without being well informed enough to have a chance to succeed at his ventures.

Kerby Jackson
Josephine County, Oregon
June 2014

GEOLOGY OF THE YREKA QUADRANGLE, SISKIYOU COUNTY, CALIFORNIA

By Preston E. Hotz

ABSTRACT

The Yreka quadrangle covers an area between the Cascade and Klamath Mountains geomorphic provinces. It includes three subdivisions of the Klamath Mountains: the eastern Klamath belt, the central metamorphic belt, and the western Paleozoic and Triassic belt.

The eastern Klamath belt is bounded on the east by Cenozoic volcanic rocks of the Cascade province and on the west by the central metamorphic belt and a belt of ultramafic rocks. The 'eastern Klamath belt includes the Gazelle Formation of Early(?) Silurian to Early Devonian age and six units of uncertain age: the Sissel Gulch Graywacke, Duzel Phyllite, Antelope Mountain Quartzite, Schulmeyer Gulch sequence, Moffett Creek Formation, and the limestone of Duzel Rock. The Sissel Gulch Graywacke is a feldspathic graywacke whose base is the Mallethead thrust. It is overlain depositionally by the Duzel Phyllite, a very uniform phyllitic calcareous siltstone and sandy siltstone. The Antelope Mountain Quartzite is composed predominantly of quartz arenite with some interbedded chert. It depositionally overlies the Duzel Phyllite and is thrust over the Schulmeyer Gulch sequence, an informal term applied to a heterogeneous unit that contains phyllite like the Duzel, quartzite like the Antelope Mountain, and discontinuous bodies of limestone. The Moffett Creek Formation is an extensive, very uniform, tectonically dismembered unit composed of blocks of fine-grained calcareous sandstone to siltstone in a matrix of pervasively sheared shale. It occurs both as a thrust sheet overlying the Duzel Phyllite and as a relatively autochthonous unit beneath the Duzel. The limestone of Duzel Rock, which contains minor amounts of interbedded chert, shale, and basalt, occurs as erosional remnants of a thrust sheet overlying the Moffett Creek Formation.

The ages of the Sissel Gulch, Duzel, and Antelope Mountain Formations, and the Schulmeyer Gulch sequence are unknown. They may have been broadly gradational, perhaps widely spaced lithofacies of a single stratigraphic unit before tectonic "telescoping," and the Duzel Phyllite may be a facies of the Moffett Creek Formation. The age of the Moffett Creek is uncertain but presumably is Silurian or older. The limestone of Duzel Rock is tentatively regarded as Ordovician in age.

Outcrops of the Gazelle Formation in the Yreka quadrangle are at the north end of an extensive belt of autochthonous rocks belonging to that formation. In the Yreka quadrangle the Gazelle Formation includes shale, volcanic wacke, local arkosic wacke, chert, chert-particle sandstone, and chert-pebble conglomerate. Lenticular bodies of spilite are interlayered with the sedimentary rocks. The Payton Ranch Limestone Member of the Gazelle Formation is composed of basal nodular limestone, limestone conglomerate, and local thin-bedded limestone, surmounted by thick-bedded cliff-making limestone. In the Yreka quadrangle graptolites,

1

brachiopods, corals, and conodonts indicate that the age of the Gazelle ranges from Early(?) Silurian to Early Devonian.

Ultramafic rocks, which include serpentinized peridotite, dunite and small amounts of gabbro, border the eastern Klamath belt on the north and west, probably underlie the belt, and may be an extension of the Trinity ultramafic body of Ordovician age which crops out south of the Yreka quadrangle. A terrane of greenschist, siliceous metamorphic rock, and blueschist, in which discontinuous bodies of limestone and large blocks of altered quartz diorite and trondhjemite of Ordovician(?) age occur, is exposed in a window beneath the Mallethead thrust. The terrane is interpreted as a metamorphosed tectonic melange.

The central metamorphic belt lies between sedimentary rocks of the eastern Klamath belt and serpentinite to the north and west. These rocks, which belong to the albite-epidote amphibolite facies, have a Devonian metamorphic age and are correlated with the Abrams Mica Schist of the southern Klamath Mountains.

The western Paleozoic and Triassic belt is north and west of the serpentinite belt. It includes a greenstone-chert assemblage of Permian age and the Stuart Fork Formation, a unit composed predominantly of phyllitic quartzite and numerous bodies of blueschist. The Stuart Fork Formation, which has a Triassic radiometric age, overrides the greenstone-chert assemblage on the Soap Creek Ridge thrust fault.

The Hornbrook Formation, consisting of Upper Cretaceous marine sedimentary rocks, overlaps the Schulmeyer Gulch sequence, the Stuart Fork Formation, and serpentinite in the northern part of the Yreka quadrangle.

Tertiary rocks include volcanic rocks of the Cascade province to the east of the Klamath Mountains province. The volcanic rocks include andesitic tuff breccia, some flows, and a small intrusive andesite body. Remnants of a basalt flow rest on the Gazelle Formation in the southern part of the Yreka quadrangle. Remnants of a Tertiary(?) conglomerate overlie basement rocks of the eastern Klamath terrane.

Unconsolidated Quaternary deposits include an extensive alluvial deposit in Shasta Valley that partly buries the eroded volcanic rocks, and alluvium and colluvium elsewhere along streams, in valley bottoms, and on lower slopes bordering the valleys.

The tectonic history begins in the Ordovician Period with uplift of Early Ordovician ophiolite and oceanic crust to provide detritus for Late Ordovician and Silurian deposits. Before or during Middle Silurian time, tectonism caused dismemberment of the Moffett Creek Formation and may have been responsible for low-grade regional metamorphism of the Duzel Phyllite and Sissel Gulch Graywacke. A well-defined Devonian tectonic episode is recorded in the amphibolite of the central metamorphic belt. The next tectonic event was in Early and Middle Triassic time when the Stuart Fork Formation was metamorphosed to blueschist and thrust over rocks of the western Paleozoic and Triassic belt. The age of thrusting that telescoped the Paleozoic sedimentary rocks is unknown except that it took place between Early Devonian and Late Cretaceous time. Jurassic plutonism is not recorded in the Yreka quadrangle except perhaps by gabbroic and diabasic dikes that intrude the Paleozoic rocks. Remnants of uplifted volcanic rocks document Cenozoic diastrophism in the area.

INTRODUCTION

LOCATION AND ACCESSIBILITY

The Yreka 15-minute quadrangle is in Siskiyou County, northern California (fig. 1). The coordinates of the southeast corner of

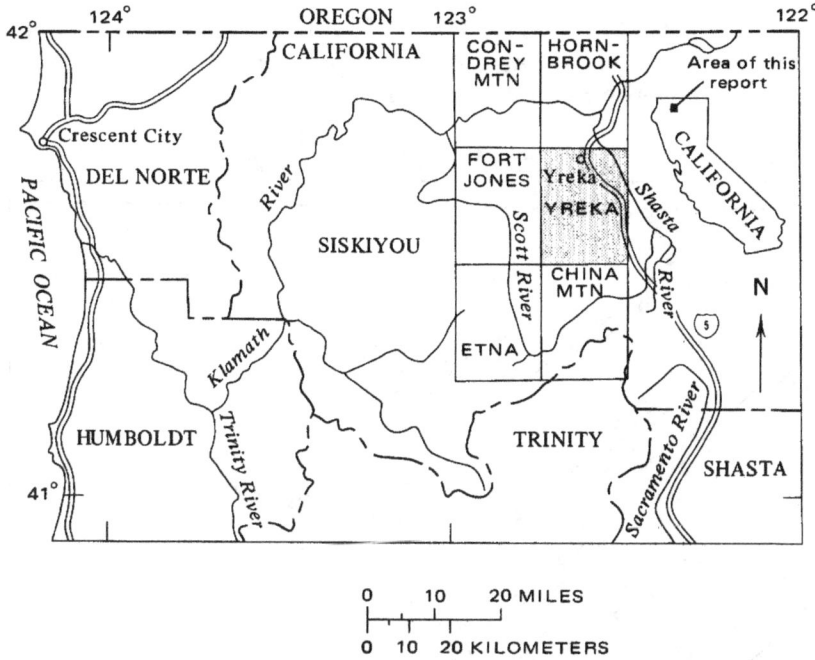

FIGURE 1.—Index map showing location of Yreka quadrangle and neighboring 15-minute quadrangles.

the quadrangle are 41°30′ N. and 122°30′ W. Yreka, the county seat and principal town, is situated on the main north-south California freeway, Interstate Route 5. The Yreka quadrangle is bordered on the north by the Hornbrook quadrangle, on the west by the Fort Jones quadrangle, on the southwest by the Etna quadrangle, and on the south by the China Mountain quadrangle.

TOPOGRAPHY

The area enclosed by the Yreka quadrangle is part of a mountainous terrain between Shasta Valley on the east and Scott Valley on the west (fig. 2). Shasta Valley is in the Cascade province; the rest of the area is in the eastern part of the Klamath Mountains province.

The altitude of Shasta Valley ranges from about 850 m (2,800 ft) near Gazelle in the southeast corner of the quadrangle to approximately 760 m (2,500 ft) in the northeast near Montague. Scott Valley is about 850 m near Etna and approximately 820 m (2,700 ft) near Fort Jones. The main north-south ridge of Antelope Mountain-Scarface Ridge attains altitudes of slightly more than 1,800 m (6,000 ft) and maintains a fairly constant level between 1,600 and 1,700 m (5,200 ft and 5,600 ft). Other summits range between about

1,200 and 1,600 m (4,000 and 5,200 ft). Slopes are moderate to steep; most valley bottoms are narrow. Two major north-flowing streams drain the area—the Shasta River in Shasta Valley and the Scott

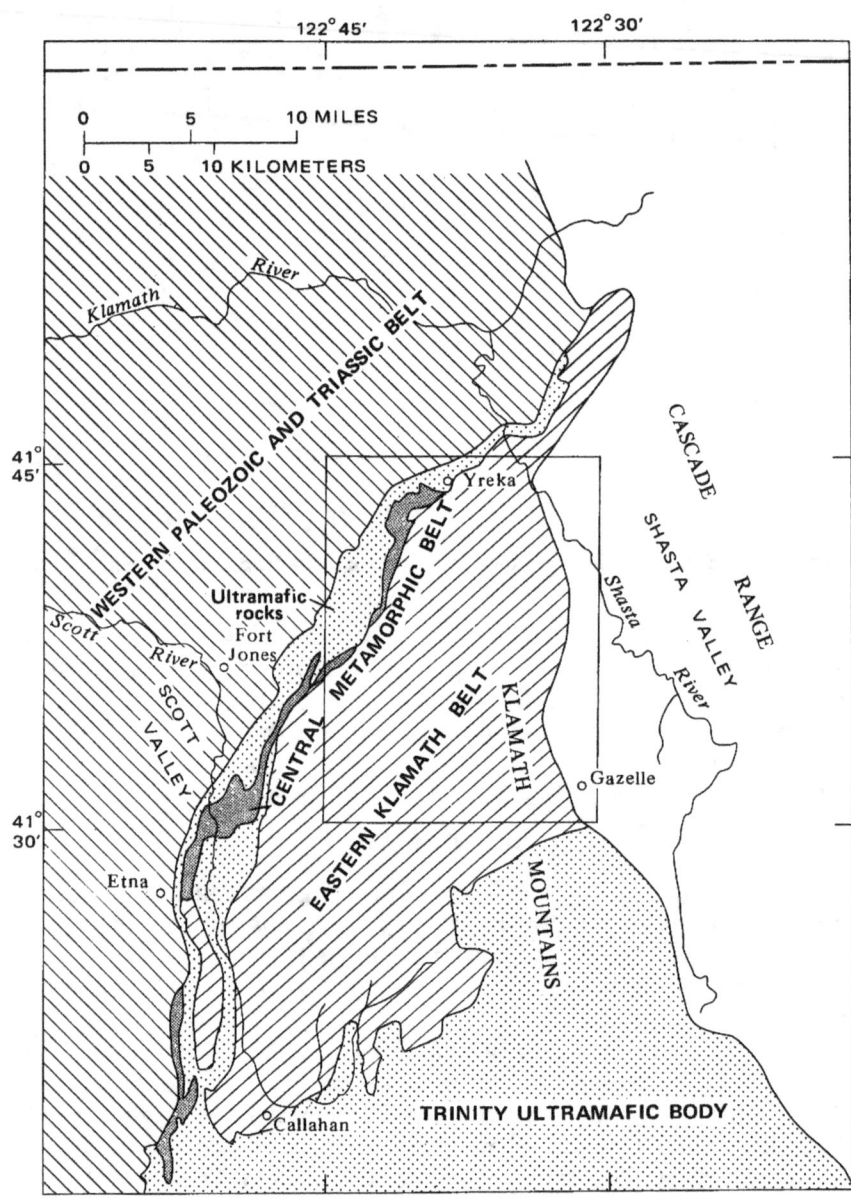

FIGURE 2.—Map showing geomorphic provinces and lithic belts in the Yreka quadrangle (outlined).

River in Scott Valley. Both streams are tributaries of the Klamath River.

The topography of Shasta Valley in the eastern part of the Yreka quadrangle is unusual. In it are numerous domical hillocks ranging in height from a few feet to as much as 75 m (250 ft). The terrain between hillocks is very flat and contains many small ponds, marshes, and alluvial flats of slow, winding streams.

PREVIOUS WORK

J. S. Diller discovered fossiliferous limestone 5 km (3 mi) southwest of Gazelle in 1884 (Diller, 1886) and 1893. He and Charles Schuchert reported on the fossils (Diller and Schuchert, 1894), which Schuchert concluded were Devonian. Merriam (1940, p. 48; 1972) subsequently restudied these and made additional collections, which he found to be of Silurian age.

Oscar Hershey's extensive traverses in the Klamath Mountains took him into the Yreka-Fort Jones area, and in 1901 (Hershey, 1901) he assigned siliceous phyllites adjacent to the serpentinite belt in the northern part of the Yreka quadrangle to the Abrams Mica Schist, a widespread formation he had described earlier in the south-central Klamath Mountains. The rocks cropping out in the range between Shasta and Scott Valleys were assigned to a unit he termed the Lower Slate Series.

Mineral deposits of the area, mainly gold, were reported by Dunn (1894), Brown (1916), Logan (1925), Averill (1931, 1935), and O'Brien (1947). Wells and Cater (1950) described some chromite deposits, and Heyl and Walker (1949) reported on limestone near Gazelle.

Howel Williams (1949) mapped the geology of Shasta Valley and the hills bordering the valley on the west from Weed northward to the California-Oregon state line. That same year the northwest corner of the Yreka and southwest part of the Hornbrook quadrangles were mapped by Masson (1949). The Cretaceous rocks in the northern part of the Yreka and Hornbrook quadrangles were included in a larger study of Upper Cretaceous rocks in southwest Oregon and northern California by Peck, Imlay, and Popenoe (1956), and by Elliott and Bostwick (1973) in the Hilt-Hornbrook area in the Hornbrook quadrangle.

Ground-water studies were made by Mack in Scott Valley (1958) and Shasta Valley (1960).

Since the 1950's, with the publication of accurate topographic base maps, more attention has been directed toward the geology of this region. Wells, Walker, and Merriam (1959) made reconnaissance studies of the terrane between Scott and Shasta Valleys from

Yreka to Callahan, including the Yreka and adjoining parts of the Fort Jones, Etna, and China Mountain quadrangles. They described the sedimentary rocks, named two formations—the Duzel Formation and the Gazelle Formation—and recognized a thrust fault between them. At about the same time, Churkin (Churkin and Langenheim, 1960) made a more detailed stratigraphic and paleontologic study of the Gazelle Formation in the Lime Gulch and Willow Creek areas, in the southern part of the Yreka and northern part of the China Mountain quadrangles, respectively, and later made the first discovery of graptolites in the Klamath Mountains (Churkin, 1965).

A gravity survey of the mountainous terrane between Scott and Shasta Valleys made by LaFehr (1966), provided data supporting the concept previously postulated by Irwin and Lipman (1962) that serpentinite underlies the Duzel and Gazelle Formations in this region.

The geochemistry and petrology of more than 70 samples collected from the Duzel and Gazelle Formations were reported by Condie and Snansieng (1971) in a study of the provenance and depositional environment of these sedimentary rocks.

In a paper on Silurian corals, Merriam (1972) discussed the stratigraphy and faunas associated with the Gazelle Formation in the Willow Creek area.

The most recent work has been in southern Yreka quadrangle, northern China Mountain quadrangle, and eastern Etna quadrangle (Potter and Boucot, 1971; Rohr and Boucot, 1971; Lindsley-Griffin, 1973; Zdanowicz, 1971).

PRESENT INVESTIGATIONS

Mapping commenced in the Yreka quadrangle in May 1969. After 6 weeks the project was recessed until June 1971 and continued during the field seasons of 1972, 1973, and part of 1975. Mapping was done on a topographic base at a scale of 1:62,500 and was carried out by means of foot traverses along roads and across country. Approximately 300 rock specimens were collected, and petrographic examinations were made of more than 200 thin sections.

GEOLOGY

REGIONAL SETTING AND GENERAL ROCK DISTRIBUTION

The Yreka quadrangle is situated at the boundary between two geomorphic provinces (fig. 2). The mountainous terrane of the Klamath Mountains province is west of the Cascade province, represented by Shasta Valley. The quadrangle embraces three geologic subprovinces of the Klamath Mountains (Irwin, 1966), including lower Paleozoic rocks of the eastern Klamath belt, rocks of the western Paleozoic and Triassic belt, and the central metamorphic belt. Rocks of the Klamath Mountains in the quadrangle also belong to either of two major subdivisions: (1) the pre-Cretaceous eugeosynclinal and plutonic "subjacent" rocks and (2) Cretaceous and younger "superjacent" rocks (Irwin, 1966).

On the east (pl. 1), Shasta Valley is occupied by Tertiary volcanic rocks partly buried by a blanket of alluvium that is apparently older than the usual Quaternary alluvium of the valley bottoms to the west. West of Shasta Valley, a wide belt of pre-Cretaceous, in part Silurian sedimentary rocks occurs in a complex structural setting.

West of the belt of sedimentary rocks is a zone of metasedimentary semischist and phyllite adjacent to a narrow belt of amphibolite that borders a belt of ultramafic rock. To the southwest, in the Fort Jones and Etna quadrangles, schistose impure marble and calcareous schists accompany the amphibolite.

West and northwest of the ultramafic rocks are highly deformed siliceous phyllite and semischist of the Stuart Fork Formation with associated abundant blue-schist. Finally, west and northwest of the phyllite and semischist is a greenstone-chert-argillite terrane. The siliceous phyllite and greenstone-chert-argillite units belong to the western Paleozoic and Triassic belt.

EASTERN KLAMATH BELT

As originally mapped by Wells, Walker, and Merriam (1959), the sedimentary rocks underlying the mountains south of Yreka between the Tertiary volcanic rocks of Shasta Valley and the belt of serpentinite to the west were assigned to two formations, the Gazelle Formation of Silurian age and, structurally overlying the Gazelle on a thrust fault, the Ordovician(?) Duzel Formation. Churkin and Langenheim (1960) mapped this thrust fault and named it the Mallethead thrust. Wells, Walker, and Merriam (1959) considered the Duzel Formation to be composed predominantly of phyllitic graywacke and relatively minor amounts of interbedded limestone and chert. Fossiliferous limestone in Horseshoe Gulch in the eastern part of the Etna quadrangle, which they believed to be part of the Duzel Formation, yielded an uppermost Ordovician fauna. Mapping by myself and by A. W. Potter (written commun., 1974) has shown, however, that the fossiliferous beds in Horseshoe Gulch are not part of the Duzel Formation.

In the Yreka quadrangle, geologic mapping of the rocks above the Mallethead thrust, previously considered to be predominantly phyllitic graywacke of the Duzel Formation (Wells and others, 1959), has demonstrated that there are five recognizable lithologic units in addition to a limestone unit at Duzel Rock that Wells, Walker, and Merriam (1959) considered part of the Duzel Formation. The lithologic units, in apparent structural succession from bottom to top above the Mallethead thrust, are (1) Sissel Gulch Graywacke, (2) Duzel Phyllite, (3) Antelope Mountain Quartzite, (4) the Schulmeyer Gulch sequence, and (5) Moffett Creek Formation. The limestone of Duzel Rock is thrust over the Moffett Creek Formation.

The sedimentary rocks of the eastern Klamath belt are described below according to their structural position beginning with the Gazelle Formation immediately below the Mallethead thrust and proceeding upward.

GAZELLE FORMATION

NAME AND DISTRIBUTION

The Gazelle Formation was named by Wells, Walker, and Merriam (1959) and also described by Churkin and Langenheim (1960) and Condie and Snansieng (1971). Merriam (1972) designated its type section as on the southwest side of Bonnet Rock on the north side of the north fork of Willow Creek in the SE¼. sec. 6, T. 42 N., R. 6 W. in the southeast part of the Yreka quadrangle. Although the Gazelle occupies a considerably larger area to the south in the China Mountain quadrangle, only the rocks in the Yreka quadrangle are discussed in this report.

The Gazelle Formation adjoins the Moffett Creek Formation on the northwest and is separated from it by a northeast-striking fault along which there are small bodies of serpentinite (pl. 1). The Gazelle is terminated on the northeast by a fault that brings it into contact with the Duzel Phyllite.

LITHOLOGY

The Gazelle Formation in Yreka quadrangle has a highly varied lithology comprising volcanic and sedimentary rocks. The sedimentary rocks include interbedded sandstone and shale, chert, chert-pebble conglomerate, and limestone. Erosional remnants of a thick limestone layer that overlies the other sedimentary and volcanic rocks forms prominent caps on several ridges. Churkin and Langenheim (1960) named this limestone the Payton Ranch Limestone Member of the Gazelle Formation.

Volcanic rocks crop out on the lower south slope of Bonnet Rock, and a large lens occurs on hill 4433. Other smaller lenticular bodies occur in the little valley west of Bonnet Rock and on ridges in the western part of sec. 6, T. 42 N., R. 6 W. The volcanic rocks west of Bonnet Rock occur in red and green shale; those on the south slope of Bonnet Rock are overlain by chert.

The volcanic rock is dark-greenish-gray rough-surfaced fine-grained to aphanitic basalt that weathers readily to a red-brown soil. In outcrops it commonly appears massive, but in places it is obviously fragmental and locally pillowed. In places fine-grained sediments are intimately mixed with the volcanic rock.

Microscopically the volcanic rock commonly has a well-preserved intersertal texture composed of ophitically intergrown laths of slightly hazy albite and clinopyroxene, with interstitial fine-grained albite and chlorite. Some of the clinopyroxene has a brownish tinge, suggesting a titaniferous composition. Opaque black metallic rods and skeletal masses, probably magnetite, usually accompanied by leucoxene, are locally present. Pumpellyite commonly accompanies some of the chlorite and may occur in veinlets with albite. Minor veinlets of carbonate may be present. In some specimens albite and augite form an intersertal texture without the ophitic intergrowth, and less commonly the rock is composed of laths of cloudy albite in a divergent intergrowth accompanied by interstitial chlorite without augite. The rock is commonly cut by narrow trains of microbreccia. These rocks are spilites of basaltic composition (table 1), and one (no. 4) has the composition of a keratophyre.

The sandstones are mainly gray to greenish-gray, fine-grained to very fine grained wackes. Characteristically, they are thick bedded and poorly stratified, although locally some thin turbidites have been observed. Individual constituents can seldom be identified megascopically.

Microscopically many of the sandstones are composed of poorly sorted angular to subrounded grains in an exceedingly fine grained matrix in which individual species can be recognized only with difficulty, or in which calcite is the predominant cement. Many specimens are volcanic wackes in which quartz grains and chert fragments are subordinate to plagioclase and rock fragments. The plagioclase is both twinned and untwinned and of sodic oligoclase to albitic composition. Commonly it is more or less sericitized and has rather hazy boundaries. Other mineral constituents are chlorite in the matrix, minor sphene, which may be in large part authigenic, gray translucent leucoxene derived by alteration of ilmenite, rarely a few grains of epidote, authigenic pyrite, and scattered grains of calcite. Lithic fragments include microcrystalline volcanic rock, shale or argillite, and minor chert.

An unbedded volcaniclastic wacke that crops out above chert and shale on the southeast side of Bonnet Rock contains megascopic clasts of mudstone and chert and some limestone. Microscopically, 95 percent of it consists of poorly sorted angular to subangular plagioclase grains ranging from 1.5 to less than 0.05 mm in a matrix of feldspar microlites and chlorite. The balance of the clasts are clinopyroxene, volcanic rock, and limestone. Irregularly shaped cavities are filled with chlorite.

A turbidite overlying the massive volcaniclastic wacke is a calcareous lithic wacke composed of angular clear chert and

TABLE 1.—*Chemical analyses of volcanic rock from the Gazelle Formation, Yreka quadrangle*

[Rapid rock analyses by Hezekiah Smith]

Sample No	1	2	3	4
Chemical analyses (weight percent)				
SiO_2	44.8	48.2	46.2	57.6
Al_2O_3	14.0	13.7	16.7	13.0
Fe_2O_3	3.1	4.2	3.4	1.5
FeO	9.1	9.4	10.8	5.9
MgO	6.0	5.1	3.3	5.4
CaO	12.3	6.9	4.6	5.9
Na_2O	3.1	4.2	4.1	4.7
K_2O	.08	.22	1.1	.09
H_2O^+	4.0	3.2	4.5	2.6
H_2O^-	.35	.50	.57	.34
TiO_2	1.5	2.8	2.6	1.2
P_2O_5	.14	.35	.35	.16
MnO	.14	.18	.21	.15
CO_2	1.3	.08	.01	.09
Sum	100	99	99	99
Norms (weight percent)				
Q	....	0.4	....	9.6
Or	.5	1.4	6.9	.6
Ab	27.0	37.1	36.9	41.4
An	25.1	18.7	21.8	14.7
Ne	.2	....	....	....
Di	23.8	11.6	....	11.5
Hy	....	17.5	16.1	16.6
Ol	11.9	....	5.0	....
Mt	4.7	6.4	5.2	2.3
Il	3.0	5.5	5.3	2.4
Ap	.35	.87	.88	.40
Cc	3.1	.19	.02	.21

Location of samples:
1. (Y-2-75) Basalt, NE¼ sec. 7, T. 42 N., R. 6 W.
2. (Y-16-75) Basalt, SW¼ NE¼ sec. 6, T. 42 N., R. 6 W.
3. (Y-27-72) Basalt, NW¼ sec. 6, T. 42 N., R. 6 W (Hill 4433)
4. (Y-17-72) Keratophyre, saddle, NE¼ sec. 12, T. 42 N., R. 7 W.

siltstone fragments and subordinate pyroxene and clasts of mudstone and limestone. The particles are loosely packed and bonded with calcite.

An unusual arkosic wacke (Williams and others, 1955, p. 292) that is significant for the insight it gives into provenance is interbedded with shale and occasional lenses of greenstone near the northeast-striking fault between the Moffett Creek and Gazelle Formations in NW¼ sec. 6, T. 42 N., R. 6 W. It is a fine-grained grayish-olive sandstone composed of angular to subrounded grains loosely packed in a chloritic matrix. Mineral grains include quartz, plagioclase, green hornblende, augite, epidote, minor green biotite, and a brown, probably chromiferous spinel. Fragments of dark fine-grained volcanic rock, keratophyre, shale or argillite, recrystallized chert or quartzite, and grains of weakly birefringent material tentatively identified as serpentine are present.

Grayish-green, grayish-red, and black shales occur as lenticular interbeds in the dominantly sandstone terrane. The shales are well indurated, and slaty fracturing is well developed locally. Lenses of thin-bedded gray to black chert are interbedded with the sandstone and shale.

Chert-pebble conglomerates, which are known to be plentiful elsewhere in the Gazelle Formation, are scantily represented here. These are massive to poorly bedded rocks composed of granules and pebbles of black and white chert, some black siliceous mudstone, and minor grains of authigenic pyrite in a porcellaneous silica matrix. Commonly the clasts are highly angular and hence the rocks are more properly called breccias.

Chert is plentiful in this part of the Gazelle Formation. It is mostly dark gray to black, but light-gray and greenish-gray varieties are also seen. Much of it is thick bedded or massive, but some is rhythmically bedded, with individual layers ranging from approximately 4 to 10 cm (2.5 to 6 in.) in thickness. At many places chert has been strongly fractured and healed by numerous crisscrossing white silica veinlets.

The Payton Ranch Limestone Member overlies sandstone and shale of the Gazelle Formation. The base of the Payton Ranch commonly is a fossiliferous rubbly weathering nodular limestone and limestone conglomerate, and locally thin-bedded limestone with shale interbeds. The basal unit is as much as 7.5 m (25 ft) thick (Merriam, 1972; Churkin and Langenheim, 1960). The thick-bedded cliff-making limestone unit that overlies the basal unit ranges from 23 to 54 m (75 to 175 ft) in thickness (Merriam, 1972; Churkin and Langenheim, 1960) and is composed of finely crystalline, fairly pure calcium carbonate. It includes some local lenses of intraformational limestone conglomerate, chert conglomerate, and gray bedded chert.

In this, the type area of the Gazelle Formation, the section is incomplete, for its base is concealed and the top is an erosion surface on the cliff-making Payton Ranch Limestone Member. Merriam (1972) assigned a total thickness of approximately 630 m (1,200 ft) to the type section, which is on the southwest side of Bonnett Rock. The thickness reported by Churkin and Langenheim (1960) for a section near here is approximately 220 m (720 ft).

Brachiopods, a trilobite, and corals obtained from unit 1, mainly from the base of the Payton Ranch Limestone Member (Churkin and Langenheim, 1960; Merriam, 1972), are of Middle or Late Silurian age. Graptolite fossils collected by Churkin (1965) in shale and siltstone about 15 m (50 ft) stratigraphically below the base of the Payton Ranch Limestone Member indicate that this part of the Gazelle Formation may be of latest Early Silurian (uppermost Llandoverian) or Middle Silurian (Wenlockian) age. Graptolites obtained in 1975 from a shale interbed in chert on the south side of Bonnet Rock (SW¼SW¼ sec. 5, T. 42 N., R. 6 W.) were identified by W. B. N. Berry (written commun., 1975). According to his report, they include "*Climacograptus* sp. probably of the *C. scalaris* group—possibly *C. scalaris normalis,* and *Climacograptus* sp.— possible *C. medius* or *rectangularis,*" probably Lower Silurian (early part of the Llandoverian Stage). The underlying volcanic rocks may be of Early Silurian or even Ordovician age. A concretion from mudstone in the upper part of the Gazelle Formation in upper Willow Creek (China Mountain quadrangle) yielded conodonts that indicate a late Siegenian or early Emsian (late Early Devonian) age (Boucot and others, 1974). The age of the Gazelle Formation is therefore considered Early(?) Silurian to Early Devonian.

SISSEL GULCH GRAYWACKE

NAME AND DISTRIBUTION

The lowermost unit above the Mallethead thrust is a graywacke here named the Sissel Gulch Graywacke for exposures in Sissel Gulch, a tributary to Moffett Creek in the southwest part of the Yreka quadrangle. The upper Moffett Creek drainage, including Sissel Gulch, from the SW¼ sec. 32, T. 43 N., R. 7 W., to the south boundary of the quadrangle in sec. 7, T. 42 N., R. 7 W., is designated the type locality.

The formation is well exposed in a quarry in Moffett Creek near the mouth of Sissel Gulch. The unit also occurs in the northeastern part of the Etna quadrangle in upper McConaughy Gulch. In all these localities, the graywacke overlies greenschist on a thrust fault; in upper Lime Gulch it is thrust over the Moffett Creek Formation.

LITHOLOGY

Graywacke beds that range from approximately 15 cm (6 in.) to 1 m (3 ft) in thickness are separated by phyllitic shale partings 3 cm (1 in.) or less thick. Commonly the shaly interbeds are distorted by squeezing between the more competent graywacke beds. Fresh graywacke is greenish gray, fine grained, and hard. Slight variations in grain size and, in places, graded bedding can be seen. A faint parting plane parallel to the bedding is commonly visible. In part, the planar structure is an incipient cleavage, but in places it probably is of sedimentary origin. Commonly there are some thin (3 to 1.5 mm, 0.12 to 0.06 in.) veinlets parallel to the bedding.

Grains range from about 0.05 to 3 mm (0.002 to 0.12 in.) in greatest dimension. Sorting is poor, and the fragments are angular to subangular, although a few subrounded clasts are present. Commonly the grains have their greatest dimensions subparallel to one another, giving the texture a slight preferred orientation. The fragments are loosely packed in a microscopic to submicroscopic matrix.

The rock is so fine grained that under the hand lens only a few grains of feldspar, quartz, and rock fragments can be recognized. Under the microscope, however, these are clearly the principal constituents. Quartz is most plentiful. Plagioclase amounts to approximately 10 percent. It is uniformly fresh, with clearly visible polysynthetic twinning. Twinning in some grains is slightly bent. The feldspar's composition, based on measurement of extinction angles, is oligoclase-andesine to andesine (An_{30-35}). Rock fragments are plentiful and include microcrystalline quartzite (probably metachert), microcrystalline siliceous volcanic rock (possible keratophyre), and less commonly very fine grained altered mafic volcanic rock. A few coarser composite grains of quartz and plagioclase are also present. No potassium feldspar was identified either in thin section or by etching and staining ground surfaces of rock specimens. Minor constituents identified include mainly sphene and also epidote, apatite, and chlorite. One grain of garnet was seen.

The matrix is exceedingly fine grained, and individual constituents cannot be clearly resolved, but it appears to be composed of the same material as the larger fragments plus pale chlorite and sericite. In some specimens the matrix is obviously recrystallized and wraps around the larger grains, whose boundaries with the matrix are hazy due to reaction. The graywacke is conspicuously free of carbonate, either as clasts or matrix.

The rock is classified as a feldspathic graywacke. The chemical analysis of sample 1 in table 2 shows its andesitic composition.

STRATIGRAPHY AND THICKNESS

Wherever it occurs, the Sissel Gulch Graywacke underlies the Duzel Phyllite. The nature of the contact is uncertain because it is not exposed, but the apparently conformable dip of the bedding in the two units and apparent interfingering in the upper part of Sissel Gulch suggest that the Duzel Phyllite is deposited on the graywacke. Estimates of thickness have little meaning because the lower contact of the graywacke unit is faulted, but the exposed thickness in Sissel Gulch and the upper part of Moffett Creek is 600 to 700 m (2,000 to 2,500 ft).

DUZEL PHYLLITE

NAME AND DISTRIBUTION

The Duzel Formation was named by Wells, Walker, and Merriam (1959) for exposures in Duzel Creek in the southwestern part of the Yreka quadrangle. The unit is herein renamed the Duzel Phyllite because of its predominantly phyllitic lithology. It crops out on Scarface Ridge and in the valleys of Moffett Creek and the upper part of Duzel Creek, in the hills north of Bonnet Rock and flanking the west side of Shasta Valley and extends southeastward into the eastern part of the Fort Jones and Etna quadrangles. The type locality is on Scarface Ridge along the county road in secs. 22, 23, and SE¼ sec. 14, T. 43 N., R. 7 W.

LITHOLOGY

A distinguishing characteristic of the Duzel Phyllite is its great uniformity. Phyllitic calcareous siltstone and calcareous sandy siltstone are the predominant rocks, excepting a few local lenses of quartz arenite and chert a few meters wide and a few hundred meters in length in the eastern part of the unit north of Long Gulch on the flank of Antelope Mountain and in the hills east of Cram Gulch. Some of the larger lenses, which are several hundred meters long, are shown on the geologic map (pl. 1).

Where completely fresh the phyllitic siltstone is light to medium gray. Fractures across the cleavage and bedding of slightly weathered specimens are pale yellowish brown. A characteristic feature of the siltstone is its calcareous composition. Most specimens effervesce vigorously with dilute hydrochloric acid. The

TABLE 2.— *Chemical and semiquantitative spectrographic analyses of sedimentary rocks, Yreka quadrangle*

[Rapid rock analyses by Paul Elmore. Semiquantitative spectrographic analyses by Chris Heroupolos]

Sample No	1	2	3	4
Chemical analyses (weight percent)				
SiO_2	66.3	54.4	63.6	68.4
Al_2O_3	15.5	8.4	8.2	8.5
Fe_2O_3	1.0	1.4	.78	1.1
FeO	3.0	1.6	2.2	1.9
MgO	2.0	2.0	2.4	2.4
CaO	3.3	14.3	8.7	6.1
Na_2O	4.4	.95	1.5	1.1
K_2O	1.2	1.6	1.6	1.9
H_2O^+	2.4	1.8	1.3	1.7
H_2O^-	.07	.15	.08	.13
TiO_2	.59	.44	.54	.49
P_2O_5	.15	.10	.08	.10
MnO	.06	.04	.06	.01
CO_2	.06	12.7	9.0	6.6
Sum	100	100	100	100
Specific gravity	2.72	2.65	2.68	2.68
Semiquantitative spectrographic analyses (parts per million)				
B	N	30	30	20
Ba	200	200	200	300
Co	7	5	5	7
Cr	100	70	50	50
Cu	30	20	15	20
Ni	30	20	15	15
Pb	N	10	15	10
Sc	15	7	7	7
Sr	150	200	150	150
V	70	30	30	30
Y	10	15	20	15
Zr	70	100	150	100
Ga	15	10	10	10
Yb	1.5	1.5	1.5	1.5

Sample description and location:
1. Feldspathic wacke, Sissel Gulch Graywacke (Y-36-39). SW¼ sec. 32, T. 43 N., R. 7 W., Yreka quadrangle.
2. Calcareous sandy siltstone, Duzel Formation (Y-23-69). Ridge east of Cram Gulch, SW¼ sec. 8, T. 43 N., R. 6 W., Yreka quadrangle.
3. Calcareous phyllite, Duzel Formation (Y-17-69). Duzel Creek, south edge sec. 23, T. 43 N., R. 8 W., Yreka quadrangle.
4. Quartz wacke, Moffett Creek Formation (Y-12-69). SE¼NE¼ sec. 19, T. 44 N., R. 7 W., Yreka quadrangle.

phyllite is thinly layered at 1- to 2-mm intervals and is fissile. The cleavage apparently parallels the bedding. Commonly the rock is intensely crumpled and folded on a small scale.

. The constituents of the siltstone cannot be identified megascopically, although thin elongate dark and light grains and lamellae are visible. Measured under the microscope the clasts range in size from approximately 0.025 mm to as much as 0.8 mm; most commonly, however, they average about 0.1 mm. The principal grains are quartz, plagioclase, and colorless mica. The quartz and plagioclase grains are angular to subangular; many quartz fragments are elongate splinters. In some specimens the quartz and plagioclase have irregular to cuspate boundaries. Thin flakes of muscovite are molded against quartz and feldspar grains. A few squarish to subrounded grains of sphene and colorless unidentified strongly birefringent highly refractive grains, plus a few black metallic grains, constitute the balance of the clasts. Scattered grains of pyrite are also probably authigenic.

The matrix is very fine grained (0.005 to 0.01 mm), apparently recrystallized anhedral quartz, plagioclase, very fine shreddy muscovite along cleavages, and abundant calcite. The calcite is in irregular shapes and lensoid particles parallel to the foliation; some may be clastic, but all of it appears to be crystalloblastic. An incipient cleavage is apparent with colorless mica and gray opaque material concentrated in thin, somewhat discontinuous folia.

The phyllitic siltstone has the composition of a calcareous feldspathic graywacke. The chemical composition of a representative specimen is presented in no. 3, table 2.

In the hills bordering Shasta Valley east of Cram Gulch and north of Grenada the rocks are sandier and less phyllitic. Slopes underlain by this facies characteristically have abundant platy fragments of brown-weathering fine-grained siltstone. In the rare outcrops the rock is a well-bedded brown-weathering siltstone that is medium gray on fresh surfaces. Slightly phyllitic shale is interbedded. Weathered surfaces have a sandy appearance because the rock is calcareous and weathering accentuates the granularity.

This rock is a calcareous sandy siltstone composed of angular grains about 0.1 mm in diameter, loosely packed in a microscopic matrix. The recognizable grains are mostly quartz. Plagioclase is rare; flakes of muscovite are common. Small rhomboid grains of brown-rimmed, probably authigenic dolomite are plentiful. The cementing material is abundant colorless calcite and, as shown by X-ray pattern, microscopic to submicroscopic quartz, chlorite, and muscovite. No indication of incipient cleavage or recrystallization is apparent in these rocks. The chemical composition of a specimen of sandy siltstone is presented in no. 2, table 2.

STRATIGRAPHY

The Duzel Phyllite overlies the Sissel Gulch Graywacke, apparently depositionally, but it is in thrust-fault contact over the Gazelle Formation, the Schulmeyer Gulch sequence, and greenschist in the southern part of the Yreka quadrangle. Its thickness is unknown because of structural complications.

The age of the Duzel Phyllite is unknown. Fossils have not been found in it, and fossiliferous strata in Horseshoe Gulch (Etna quadrangle) that Wells, Walker, and Merriam (1959) believed to be part of their Duzel Formation are below a thrust plate of the Duzel Phyllite. In parts of the Yreka quadrangle the Duzel Phyllite is thrust over the Moffett Creek Formation. The thrust plate may also have overlain the fossiliferous Gazelle Formation but has been removed by erosion. The calcareous sandy siltstone in the hills east of Cram Gulch resembles the brown-weathering calcareous wacke of the Moffett Creek Formation megascopically, and both rocks have similar mineral compositions, including abundant muscovite and authigenic dolomite. Their chemical compositions are also very similar (table 2). The similarities may indicate that the Duzel Phyllite is a facies of the Moffett Creek Formation, which is Silurian or older.

ANTELOPE MOUNTAIN QUARTZITE

NAME AND DISTRIBUTION

The Antelope Mountain Quartzite is a distinctive lithologic unit here named for Antelope Mountain in the central part of the Yreka quadrangle. Its type locality is the main ridge of Antelope Mountain in secs. 26 and 35, T. 44 N., R. 7 W. and secs 2, 3, and 10, T. 43. N., R. 7 W., where it is well exposed. It structurally overlies the Duzel Phyllite and the Schulmeyer Gulch sequence; it is overlain by the Cretaceous Hornbrook Formation, Tertiary volcanic rocks, and Quaternary alluvium.

The Antelope Mountain Quartzite has a maximum outcrop width of about 5.5 km (3.5 mi) and extends east from Antelope Mountain to the low hills east of Cram Gulch. It extends north to the alluviated valley through which the main highway passes and crops out again in the low hills northwest of Grenada. The northeast-southwest outcrop length is approximately 14.5 km (9 mi). A cliffy topography is commonly developed by erosion in areas underlain by these rocks.

LITHOLOGY

The Antelope Mountain Quartzite is predominantly quartz arenite, but beds and lenses of chert alternate with the clastic rocks, mainly in a zone approximately 1 to 1.5 km (0.5 to 0.75 mi) wide adjacent to the contact with the Duzel Phyllite.

In outcrops the rocks are well bedded, commonly from 0.5 to 1.5 m (2 to 5 ft) thick, rarely as much as 3 m (10 ft), and in places they are thin and rhythmically bedded. Locally they show graded bedding. Interbeds are tan to greenish-brown and maroon shale. The quartzitic beds are mainly light colored, commonly greenish gray and also gray, brown, and tan. A distinctive but less common variety is grayish red. Commonly the quartz arenite has an incipient to well-developed parting parallel to the bedding. Some specimens, however, are massive, and fractures break across the grains.

The quartz arenites are medium to coarse grained and moderately well to poorly sorted. Grains range from about 0.25 to 4 mm and most commonly from 1 to 2 mm in diameter. In general, they are loosely packed in a very fine (0.005 to 0.02 mm) siliceous matrix. The clasts range from angular to rounded but most commonly are subrounded to rounded. Some beds are composed of clasts ranging in size from granules to small pebbles.

The predominant (more than 90 percent) clasts are quartz (fig. 3). Under the microscope many quartz grains show strain shadows. Some are composite orthoquartzites; a few are fine-grained recrystallized chert. Less than 5 percent of the clasts are feldspar, mainly fresh plagioclase, and a few are potassium feldspar, including microcline. A few specimens containing more than 5 percent feldspar (5 to 6 percent) are classed as feldspathic arenites. Flakes of muscovite are rare, and traces of heavy minerals, including zircon, sphene, tourmaline, and altered ilmeno-magnetite, are present. The very fine grained to silty matrix is composed of quartz, colorless mica, chlorite, and in some specimens pale biotite. The reddish arenites have abundant very finely divided hematite in the matrix. The matrix is recrystallized in many specimens.

STRATIGRAPHY AND THICKNESS

The Antelope Mountain Quartzite depositionally overlies the Duzel Phyllite east of Cram Gulch and northeast of Pythian Cave. At most places, however, the contact is modified by faulting. On the west side of Antelope Mountain, the quartzite is thrust over the Schulmeyer Gulch sequence and the Duzel Phyllite.

Lenses of chert and quartzite identical to rocks in the Antelope

Mountain Quartzite occur in the Duzel Phyllite southeast of Antelope Mountain, and at the north end of Guys Gulch and Cram Gulch similar phyllite is interbedded with chert and quartzite of the Antelope Mountain Quartzite. The Antelope Mountain Quartzite may be a facies of the Duzel Phyllite. It is tentatively regarded as of Ordovician(?) age.

There is some folding of beds but apparently not the highly complex deformation that has taken place in the less competent phyllitic rocks. The predominant large-scale structural feature is a north-plunging synclinal fold whose axis is approximately coincident with the northeast-trending ridge of Antelope Mountain. There may be as much as 1,500 to 1,700 m (4,900 to 5,500 ft) of strata.

SCHULMEYER GULCH SEQUENCE

NAME AND DISTRIBUTION

The term Schulmeyer Gulch sequence is informally used here for a heterogeneous unit that underlies an area west of Antelope Mountain and extends from Moffett Creek northward into the low hills southeast of Yreka. Southwest of Yreka the unit also occupies a narrow belt adjacent to amphibolite or serpentinite on the west and the Moffett Creek Formation and Duzel Phyllite on the east.

LITHOLOGY

A characteristic feature of this unit is its heterogeneity. It contains phyllitic siltstone, some of which is calcareous and indistinguishable from the Duzel Phyllite, and beds and lenses of quartz arenite and chert identical to rocks in the Antelope Mountain Quartzite. Distinguishing it from these two units, however, are discontinuous bodies of limestone. The limestone amounts to approximately 25 percent of the exposed area of the unit.

Quartz arenite and chert are more plentiful than phyllite. Some of the quartz arenite is like that in the Antelope Mountain quartzite, but more commonly it is phyllitic and tends to be thin bedded. The chert is microcrystalline, well bedded, white, gray, or black. Although some of the phyllite is calcareous like the Duzel Phyllite, more commonly it is greenish-gray or brown to maroon noncalcareous phyllite or shale. Chert and shale or phyllite are associated with limestone in this unit, but quartz arenite and limestone tend not to occur together. The limestones, which are gray to black, fairly pure, and finely crystalline, are apparently unfossiliferous, although no extensive search for fossils has been made. The limestone bodies range from a few meters wide and a few tens of meters long, to more than 600 m (2,000 ft) wide and 3 km (2 mi) long.

A small amount of metavolcanic rock with interlayered lime-

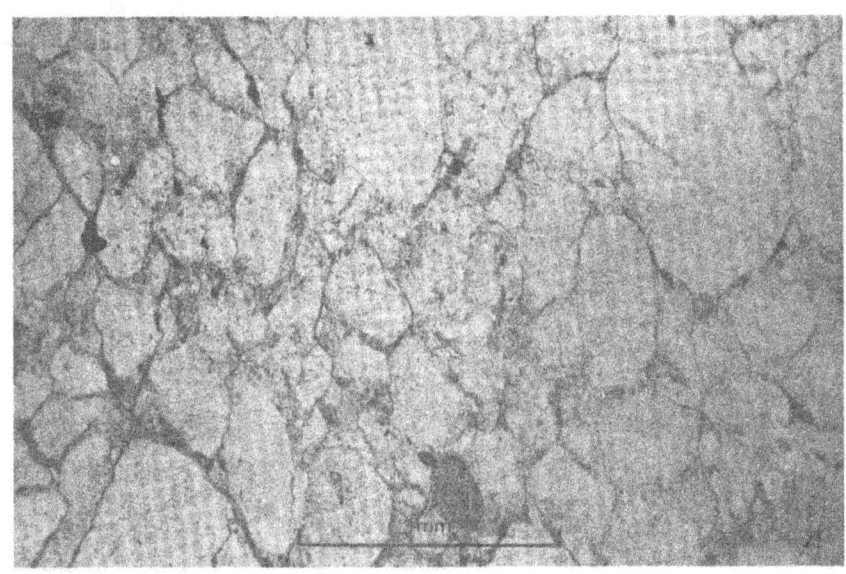

FIGURE 3.—Photomicrographs of Antelope Mountain Quartzite. *A*, Closely packed quartz grains with interstitial silica cement and secondary white mica and chlorite. *B*, Loosely packed angular to subangular quartz grains. Note strain

stone and quartzite occurs in the Schulmeyer Gulch sequence on the west side of the low ridge east of the highway in Yreka and at the east end of the ridge north of the mouth of Greenhorn Creek. The rock is pale greenish gray and very fine grained and ranges

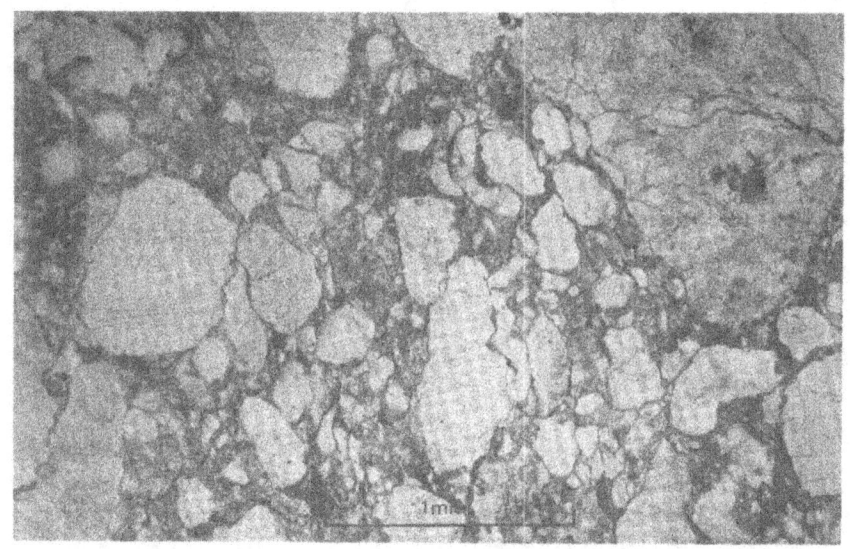

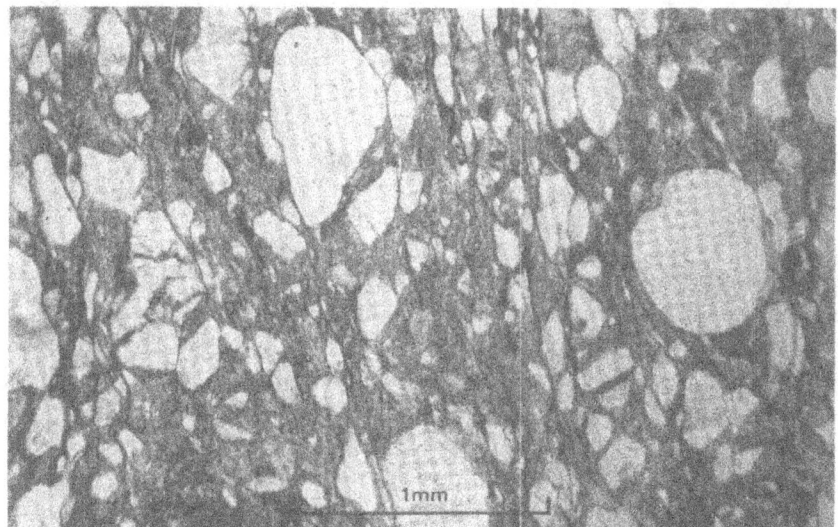

shadows in some. Crossed polarizers. *C*, Poorly sorted angular to subangular quartz grains. Abundant hematite dust in the matrix. *D*, Subangular to rounded quartz grains in matrix that shows incipient cleavage.

from phyllitic to massive. The massive variety has some small squarish chlorite pseudomorphs, probably after pyroxene, and some chlorite amygdules. The phyllitic variety is composed of microscopic pale chlorite with scattered dusty sphene and minor epidote. Very fine grained quartz and albite are interlaminated with the chlorite. The massive variety contains abundant very pale green actinolite in addition to chlorite. Although these rocks are on strike with the amphibolite to the west, they are separated by alluvium along Yreka Creek. They are unlike the amphibolite, which is of higher metamorphic grade and completely recrystallized.

Semischists adjacent to amphibolite.—In a narrow discontinuous zone adjacent to the amphibolite belt, some rocks of the Schulmeyer Gulch sequence are metamorphosed to semischists. The rocks affected are mainly the siltstone and quartz arenite, although, in places, chert shows the effects of dynamic recrystallization.

A common variety is a microcrystalline finely laminated metasiltstone composed of quartz and plagioclase porphyroclasts in a mosaic of partly recrystallized quartz, albite, and commonly shreddy chlorite. The scattered porphyroclasts are 0.1 to 0.5 mm in greatest dimension; the matrix grain size ranges from 0.01 mm or less to about 0.06 mm. Other common constituents are pale epidote or clinozoisite and colorless mica, less commonly pale-brown biotite, minor chlorite, and fine acicular actinolite.

The foliation is the result of very fine compositional layering parallel to parting surfaces or incipient cleavage. Epidote-clinozoisite granules, mica, and chlorite are concentrated in folia that are discontinuous in some specimens and in others are continuous and appear to represent original sedimentary layering. These rocks are not calcareous, but some specimens have very narrow lamina of calcite. Scattered very small rounded grains of sphene probably are original clastic grains.

Metamorphosed quartz arenites are very fine grained to microcrystalline with faint to fairly well developed partings. Individual constituents are megascopically invisible in some specimens; in others, quartz grains in a microscopic matrix can be recognized. The microscope shows (fig. 4) the rock to be commonly composed of subangular to subrounded clasts in a microscopic groundmass in which cleavage ranges from widely spaced parallel incipient parting surfaces to fairly well developed closely spaced cleavage. The clasts are predominantly quartz, mostly individual grains, but some are composite crystalloblastic quartz. Fragments of microcrystalline metachert may be present but are not common, and grains of plagioclase are present but very rare. The clasts range from about 0.1 to as much as 2.5 mm in greatest dimension; most grains are about 0.5 mm long. Many of the clasts are elongate and have a moderate to well-developed preferred orientation with their

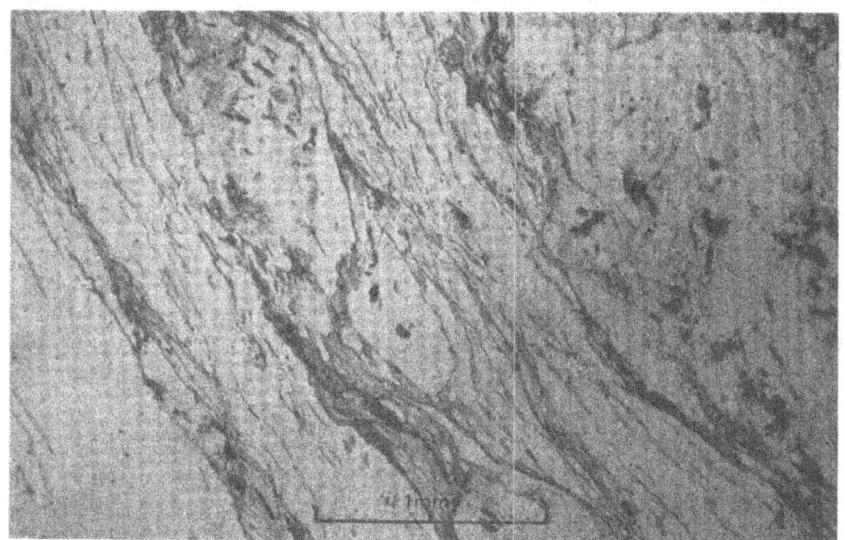

FIGURE 4.—Photomicrographs of semischist adjacent to amphibolite in Schulmeyer Gulch sequence. A, Quartzofeldspathic semischist. B, Same specimen, crossed polarizers.

long axes parallel to parting and cleavage in the matrix. In specimens with the best cleavage, many of the quartz grains are spindle shaped. Grain boundaries range from smooth to cuspate. The quartz commonly shows strongly developed strain shadows, and some of the grains have throughgoing microscopic recrystallized "fracture trails." Trace amounts of other minerals are present, including a few muscovite flakes, small subrounded to rounded grains of apatite, sphene, and colorless unidentified, highly refractive, strongly birefringent minerals. Black opaque dust is fairly abundant in some quartz grains. Colorless porphyroblastic garnet was seen in one specimen. The matrix is very fine grained, apparently mostly crystalloblastic quartz and abundant very fine fibers of sericite, and less commonly greenish-brown biotite, which is concentrated along cleavage surfaces. The parting and cleavage surfaces are made apparent by parallel sericite flakes that commonly wrap around the larger clasts. Secondary muscovite, coarser than the sericite of the groundmass, may also be concentrated along cleavage surfaces. Metamorphosed chert in this belt is recrystallized to a microcrystalline mosaic of slightly elongate, oriented quartz grains whose boundaries are irregular to sutured. Original bedding is preserved by parallel layers of varying grain size and by wavy carbonaceous and micaceous films.

The metamorphism exhibited by these rocks does not rank higher than lower greenschist facies and may, for the most part, be more suitably classified as cataclasis. Some of the rocks may be phyllitic, but most are semischist or protomylonites.

A potassium-argon age determination of 432 m.y. (Lower Silurian) was obtained on colorless mica from quartzo-feldspathic semischist that crops out in a gully in SE¼NE¼ sec. 2., T. 43 N., R. 8 W. (M. A. Lanphere, written commun., 1977). The dated rock is structurally above amphibolite of Devonian age exposed in the lower part of the same gully.

STRATIGRAPHY

It has not been possible to recognize any consistent pattern of distribution of rocks in the area underlain by the Schulmeyer Gulch sequence. The sequence is bounded by faults, so its top and bottom are unknown. The presence of rocks characteristic of both the Antelope Mountain Quartzite and Duzel Phyllite, plus limestone, which occurs in neither, suggests that it is a sedimentary facies of the two other formations. Another explanation for the heterogeneity of the unit may be that it is a melange.

MOFFETT CREEK FORMATION

NAME AND DISTRIBUTION

The Moffett Creek Formation is here named for Moffett Creek, its type locality, where the unit is exposed in sec. 12, T. 43 N., R. 8 W. The Moffett Creek Formation occupies a belt 3 to 4 km (2 to 2.5 mi) wide in the western part of the quadrangle, where it is thrust over the Duzel Phyllite and Schulmeyer Gulch sequence. Thrust klippen of the limestone of Duzel Rock overlie the formation in the southwest corner of the quadrangle. It also crops out near Lime Gulch where it is overridden by the Sissel Gulch Graywacke and Duzel Phyllite on the Mallethead thrust fault.

LITHOLOGY

In most exposures the Moffett Creek Formation consists of blocks of sandstone or siltstone in a matrix of strongly sheared shale (fig. 5A). The blocks range in size from a few centimeters to as much as a meter long (fig. 5B, C). At some places, the sandstone beds, which range from 0.25 to 1.25 m (1 to 4 ft) in thickness, are intact, whereas the shale interbeds are much squeezed and disturbed (fig. 5D). The siltstone and sandstone commonly are massive, but in places they contain laminae or lenses of coarser sandstone. Some beds are graded and uncommonly bear flute marks. Rarely there is evidence of soft-sediment deformation.

Throughout its extent the sandy part of the formation is exceedingly uniform in appearance and composition. The rock is medium gray to dark greenish gray on fresh surfaces; weathered surfaces are moderate yellowish brown. Most specimens are so fine grained that even with a hand lens the constituents cannot be identified, except grains of quartz and flakes of mica. Commonly many small flakes of white mica are visible, even on weathered surfaces, giving the rock a sparkling appearance. Microscopically the rock is composed of angular to subangular, rarely subrounded, closely packed clasts averaging about 0.1 mm and ranging from 0.25 mm (fine sand) to 0.01 mm (silt) in diameter in a very fine, submicroscopic matrix which is mostly unidentifiable as to species but in which micaceous material and calcite cement can be recognized.

The predominant clasts are quartz and minor amounts of fresh plagioclase and rare potassium feldspar. The quartz is commonly slightly strained and consists mostly of single grains. A few grains are composite, possibly derived from metamorphic rocks; a very few are recrystallized chert. Some of the quartz contains needlelike inclusions of rutile. Plagioclase is fresh, twinned, and appears to be oligoclase, around An_{30}. A few grains of potassium feldspar with

grid twinning are recognizable under the microscope. Grains of potassium feldspar are also visible on etched surfaces stained with sodium cobaltinitrite.

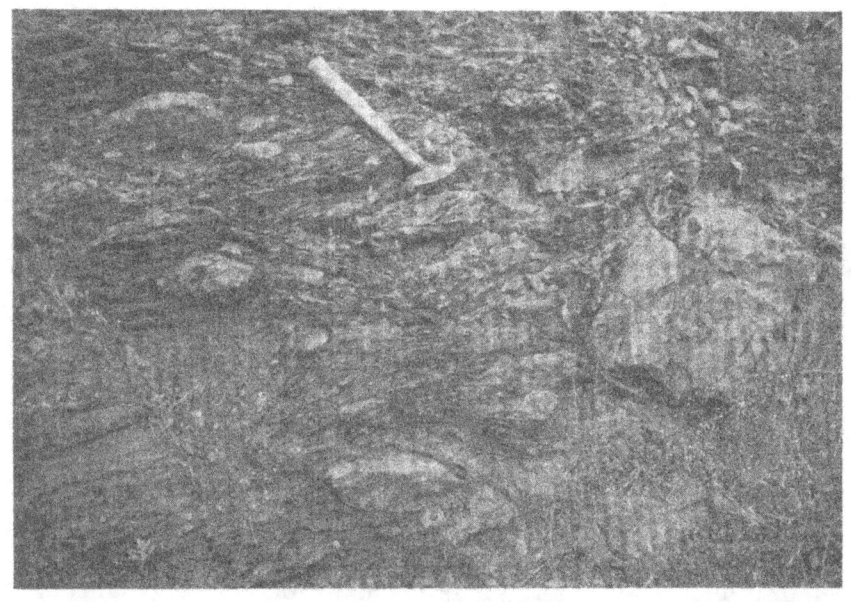

FIGURE 5.—Moffett Creek Formation. *A*, Beds of calcareous siltstone and interbeds of sheared shale. *B, C, D*, Calcareous siltstone blocks in sheared shale matrix. In *C*, note wedge-shaped outlines of some blocks.

Grains of heavy minerals are very scarce. They consist of small, rounded grains of sphene, apatite, and clear, highly refractive, strongly birefringent grains that are possibly zircon. Occasional grains of brown tourmaline are present; opaque minerals are practically absent.

Two kinds of carbonate grains occur in varying amounts: (1) single brown, occasionally diamond-shaped grains of dolomite that are probably authigenic are fairly common; (2) some specimens contain subangular grains of clear calcite that are probably clastic.

The cementing matrix is a very fine, microscopic mixture of sericitic material, minor chlorite, and hazy unidentified material. Varying amounts of calcite are present in the matrix of most specimens.

This sandstone probably can be classified as quartz wacke; possibly some specimens are feldspathic subgraywacke. Where calcite is an important part of the cement, the rocks should be called calcareous quartz wacke and calcareous feldspathic wacke. A chemical analysis of a typical specimen of the calcareous sandstone is presented in no. 4, table 2.

The Moffett Creek Formation is interrupted at a few places by isolated lenses of chert grit. These lenses are no more than a few meters thick (1 to 5 m) and a few tens of meters long. The grits range in grain size from very coarse sand (1 to 2 mm) to granule conglomerate (2 to 4 mm) and are composed of subangular to subrounded grains of chert and subrounded to rounded grains of quartz; minor amounts of plagioclase occur in some specimens. The grains are loosely to moderately well packed and are bound together by a fine sandy quartz-chert matrix with carbonate rhombs in some specimens; others have a silica cement.

Because of its widespread dismembered condition coupled with its great lithologic uniformity and apparent absence of exotic elements, the Moffett Creek Formation fits the description of a broken formation as defined by Hsü (1968). Or possibly the formation is an olistostrome, whose chaotic disruption resulted from submarine gravity sliding of unconsolidated sediments. It is difficult and may be impossible, however, to distinguish between an olistostrome and a broken formation, presumably of tectonic origin, for an olistostrome may subsequently undergo tectonic disruption. An origin by tectonic disruption is favored for the Moffett Creek Formation for several reasons, one of which is the great lateral extent of the formation—more than 45 km (28 mi)—with no observed intervals of undisturbed interbedded sandstone and shale greater than a few meters in dimension. Also, soft-sediment deformational features are rare. The occurrence of lenticular to lozenge-shaped blocks ranging from several meters to a few centimeters in size, usually bounded by smooth shear surfaces surrounded by a sheared pelitic matrix, also indicates a tectonic origin. The pelitic matrix itself, though compact, is pervasively dissected by a multitude of closely spaced, curving, braided shear surfaces and in some places, especially where weathered, the matrix resembles fault gouge.

STRATIGRAPHY AND AGE

The western belt of the Moffett Creek Formation overlies the Schulmeyer Gulch sequence and the Duzel Phyllite, whereas the eastern belt is beneath the Duzel Phyllite and Sissel Gulch Graywacke and directly below the Mallethead thrust. In the southern part of the Yreka and northern part of the China Mountain quadrangles, the Moffett Creek Formation is in contact with the Gazelle Formation along steeply dipping faults. The age of the Moffett Creek Formation is somewhat uncertain but presumably is Silurian or older, because at some places brown-weathering micaceous calcareous siltstone clasts typical of the Moffett Creek Formation occur in conglomerate of the Gazelle Formation. One such clast in mudstone of the Gazelle Formation contained Silurian brachiopods, according to A. W. Potter (written commun., 1974).

POSSIBLE FACIES RELATIONS

The possibility that the Duzel Phyllite, the Antelope Mountain Quartzite, and the Schulmeyer Gulch sequence are lithofacies of a single stratigraphic unit, and are "telescoped" by tectonic disruption, is suggested by the occurrence together of rocks characteristic of the Duzel Phyllite and Antelope Mountain Quartzite in the Schulmeyer Gulch sequence. The quartzite and chert are identical to rocks in the Antelope Mountain Quartzite, and calcareous phyllite indistinguishable from that in the Duzel Phyllite are present but not plentiful in the Schulmeyer Gulch sequence. A few lenses of chert and quartzite like those in the Antelope Mountain Quartzite are found in the Duzel Phyllite.

There may also be a facies relation between the Duzel Phyllite and the Moffett Creek Formation. This is suggested by the calcareous sandy siltstone in places in the Duzel Phyllite, which resembles the Moffett Creek's calcareous siltstone and is mineralogically identical to it. The Duzel does not, however, contain the dark shale that is so plentiful in the Moffett Creek Formation.

Mineral and rock fragments in the terrigenous clastic rocks of these units provide evidence that the original sediments had a common source area. The high content of quartz, moderate abundance of potassium feldspar and of plagioclase of intermediate composition, and common occurrence of muscovite flakes, together with the absence of fine-grained mafic rock fragments, indicate a dominantly plutonic-metamorphic provenance for these rocks.

LIMESTONE OF DUZEL ROCK

NAME AND DISTRIBUTION

The informal name limestone of Duzel Rock is used here for a group of isolated bodies on the ridge between Duzel Creek and Moffett Creek in the vicinity of the prominent peak called Duzel Rock in the southwest corner of the Yreka quadrangle. The limestone is more resistant to erosion than the underlying Moffett Creek Formation and stands out boldly above the surrounding more subdued terrane. The largest body, including the Duzel Rock, occupies a subcircular area of about 1.5 km (1 mi) maximum diameter. Less than 200 m (0.1 mi) to the south is a club-shaped body about 1.2 km (0.7 mi) long and 250 m to 1 km (0.2 to 0.6 mi) wide. A body of limestone barely 400 m (1,300 ft) in greatest dimension crops out approximately 250 m (800 ft) below Duzel Rock, on the west slope of Moffett Creek.

The recognition of these masses as erosional remnants (klippen) of a thrust sheet depends entirely upon structural data, for paleontologic data are lacking and stratigraphic evidence of the age of the rock is at best vague or conjectural.

The trace of the contact in relation to the topography at Duzel Rock and the ridge to the south is that of a smooth, flat surface that is nearly horizontal to gently inclined westward to northwestward. The beds strike north to northwest, dip eastward, and terminate abruptly at the contact with the underlying Moffett Creek Formation. Where the contact is exposed on the east side of Duzel Rock and along the road to the southwest, rocks above and below the contact are sheared, and water seepages and springs occur at a few places along the contact. A conspicuous zone of white quartz, probably a replacement of brecciated limestone, occupies the southern contact of the Duzel Rock klippe.

LITHOLOGY

These klippen are predominantly limestone with minor amounts of interbedded chert, shale, and volcanic rock. Some of the limestone is brown weathering and sandy. Some is pure, dark gray and brown weathering, and some, also pure, is gray and weathers light gray. Some is cherty. In general, the limestone is well bedded, ranging from thin and platy to medium bedded, but some—for example the unit that constitutes the great bluff of Duzel Rock—is very thick bedded to massive, and oolitic. None of the limestone is fossiliferous, although some is fetid when freshly broken, indicating the former presence of organic material. Coarse limestone breccias are present locally, for example on the gentle slopes of the southwest lobe of the Duzel Rock klippe, and in part of the small eastern klippe. The breccias are composed of angular fragments as much as a meter (3 ft) across firmly bonded by a limestone cement. Crude bedding can be seen in places. All the limestones are recrystallized, which may account for the apparent lack of fossils, although ooliths are preserved in some recrystallized specimens.

In the Duzel Rock klippe and the one to the southwest, some basaltic volcanic rock is interbedded with the sedimentary rocks in lenticular beds up to about 60 m (200 ft) thick. A bed in the Duzel Rock klippe, well exposed in bluffs on the steep south side of the ridge, is composed of chocolate-brown-weathering dusky-red volcanic breccia and pillow lava. Aphanitic and commonly amygdular fragments, as much as 30 cm (1 ft) in diameter, are enclosed in a fine-grained greenish fragmental volcanic and carbonate matrix. A few fragments of limestone are enclosed in the matrix, and some brownish limestone is interbedded with the volcanic rock. The volcanic rock in the klippe south of Duzel Rock is also fragmental and amygdular.

All these volcanic rocks are altered. Although their original textures and structures are preserved, their minerals are mostly secondary. The albitic plagioclase is mostly hazy and saussuritized, although relicts of labradorite phenocrysts are preserved in a specimen from volcanic breccia of the Duzel Rock klippe. This specimen also retains phenocrysts of unaltered augite, but serpentine forms pseudomorphs after the original olivine. Most commonly, besides saussuritized plagioclase, the principal constituents are interstitial masses of chlorophaeite, abundant rods and skeletal grids of magnetite, and interstitial gray translucent leucoxene, plus amygdules and veinlets of calcite. An analyzed clast (table 3) from the volcanic breccia is a basalt rich in TiO_2, which is reflected by titanite (sphene) in the norm.

AGE

The age of the limestone of Duzel Rock is unknown. A tentative correlation is made with a limestone klippe at Facey Rock, 16 km (10 mi) to the south-southwest where the rocks and structure are similar and the limestone has yielded Ordovician conodonts (Porter, 1973). The limestone of Duzel Rock is therefore assigned an Ordovician(?) age.

PROVENANCE AND CONDITIONS OF DEPOSITION OF ROCKS IN THE EASTERN KLAMATH BELT

Mineralogical and chemical data (Condie and Snansieng, 1971), as well as overall lithology, indicate that rocks above the Mallethead thrust—the Sissel Gulch Graywacke, Duzel Phyllite, Antelope Mountain Quartzite, and Moffett Creek Formation—and the Gazelle Formation below the thrust had different source areas and depositional environments.

TABLE 3.—*Chemical analysis of volcanic rock from Duzel Rock*
[Rapid rock analysis by Hezekiah Smith]

Sample No	1
Chemical analysis (weight percent)	
SiO_2	45.2
Al_2O_3	12.6
Fe_2O_3	11.7
FeO	2.2
MgO	4.2
CaO	8.4
Na_2O	4.0
K_2O	0.60
H_2O^+	2.2
H_2O^-	1.1
TiO_2	3.2
P_2O_5	0.49
MnO	0.12
CO_2	2.6
Sum	99
Norm (weight percent)	
Q	5.7
Or	3.7
Ab	35.1
An	15.2
Di	4.2
Hy	8.9
Ol	....
Mt	....
Hm	12.1
Il	5.1
Tn	1.6
Ap	1.2
Cc	6.1

Location of sample:
1. (Y-14-71) Clast in volcanic breccia, south end of Duzel Rock. NW¼ sec. 1, T. 42 N., R. 8 W.

The mineralogy of the Sissel Gulch Graywacke is indicative of a volcanic source, possibly of intermediate composition, but the plentiful quartz content and occasional composite grains of quartz and plagioclase suggest contributions from a more silicic, possibly plutonic terrane. In the Duzel Phyllite, Antelope Mountain Quartzite, and Moffett Creek Formation the predominance of quartz over plagioclase, the occasional occurrence of potassium feldspar, and the ubiquitous presence of muscovite flakes, together with minor amounts of sphene, apatite, zircon, and tourmaline strongly suggest a sialic continental source for the sediments or, alternatively, plutonic and metamorphic rocks exposed in deeply eroded parts of a volcanic arc.

In contrast, most of the wackes of the Gazelle Formation probably were derived from volcanic rocks that were being erupted nearby, either subaerially or on the sea floor, during deposition of the sedimentary rocks. The plagioclase in the wackes is albite, which Churkin and Langenheim (1960) suggest is formed by diagenetic alteration of more calcic plagioclase. However, it seems unlikely that a plagioclase as calcic as andesine changes to albite in unmetamorphosed rocks such as these. More probably the albite is original and was derived from spilitic basalt or keratophyre and altered quartz diorite or trondhjemite. It is impossible to tell the composition of the volcanic rock fragments that occur in the sandstone except that they are not siliceous and may range from spilite to andesite or basalt.

Other constituents of the sandstones, including brown spinel, possibly serpentine, hornblende, augite, epidote, and quartz, are strong evidence that the sediments were derived from an area in which serpentinite and probably gabbro or diabase and quartz diorite or trondhjemite were exposed to erosion.

The contrasts in provenance are also illustrated by the chemical data. Sample 1 in table 2, from the Sissel Gulch Graywacke with its relatively high alkali content, predominance of Na_2O over K_2O, and high Al_2O_3 content, reflects a volcanic source of andesitic composition. Samples 2, 3, and 4 (table 1) from the Duzel Phyllite and Moffett Creek Formation contain relatively little Al_2O_3 and total alkalis are low, with K_2O ranging from nearly equal to definitely greater than Na_2O. This is in contrast to most of the graywackes from the Gazelle Formation sampled by Condie and Snansieng (1971, table 2, p. 744) in which Na_2O content is higher than the K_2O content, total iron (as Fe_2O_3) is high, and Al_2O_3 is moderate.

The Sissel Gulch Graywacke and especially the Duzel Phyllite and Moffett Creek Formation have great lithologic uniformity both laterally and apparently vertically. Their uniformly fine-grained character is indicative of a relatively quiet, moderately deep water depositional environment in a subsiding marine basin. The Antelope Mountain Quartzite with its coarser, moderately to well-rounded grains and great purity was deposited near shore in shallow water where wave action was an effective washing and rounding agent. The occurrence of maroon shaly and silty inter-beds and a few strata of hematite-cemented quartzite further attest to a temporary, locally subaerial environment. Its restricted occurrence and apparent continuity with the Duzel Phyllite suggests that it was laid down locally on the border of the basin in which the Duzel accumulated.

The Gazelle Formation in the Yreka quadrangle commonly shows great lateral and vertical heterogeneity, although at some places the units have considerable lateral continuity. Much of the heterogeneity observed is due to widespread faulting and dismemberment that in places is severe enough for the unit to be classed as a tectonic melange or broken formation. It seems likely, however, that the rocks were deposited on a complex topography with locally contrasting environments, thus producing abrupt lateral facies changes and much more heterogeneity. Discontinuous lenses of bedded chert and limestone were deposited locally in temporary basins of relatively quiet water. Unstable slopes gave rise to density and turbidity currents, debris flows, and slide blocks. The thick ridge-capping Payton Ranch Limestone Member, however, represents the establishment of a more permanent, relatively extensive site favorable for the deposition of calcium carbonate.

A reef origin has long been suspected for the thick, craggy limestone bodies of the Payton Ranch Limestone Member, but reefal structures have not been seen. Much of the coarser textured carbonate is biogenic, but the origin of the fine-grained pure limestone facies is unexplained. The limestone could represent accumulations of bioclastic material and lime mud on local shallow banks in protected embayments that were bypassed by currents transporting the coarser clastic material that formed the wackes (Merriam, 1972; Churkin and Langenheim, 1960).

ULTRAMAFIC ROCKS AND GABBRO

One of the outstanding geologic features of the area is the continuous belt of ultramafic rocks that extends from the Hornbrook quadrangle through the northwest part of the Yreka quadrangle into the Fort Jones and Etna quadrangles. Near Callahan the belt may connect with ultramafic rocks of the very extensive

Trinity ultramafic body (fig. 2). In the Yreka quadrangle the belt ranges in width from a few tens of meters at its north end to as much as 4 km (2.5 mi) at the boundary between the Fort Jones and Yreka quadrangles.

The ultramafic rocks are serpentinized peridotite and dunite and range in structure from massive to highly sheared and broken. Some of the massive rock is incompletely serpentinized and contains relict olivine and diallage in a groundmass of serpentine minerals. The highly sheared rock is completely serpentinized. The predominant rock is serpentinized harzburgite, with minor amounts of serpentinized dunite. In places, for example the ridge in SW¼ sec. 6 and NE¼ sec. 7, T. 44 N., R. 7 W., a prominent planar structure, probably of tectonic origin, has been developed in dunite. The dunite is only partly serpentinized (about 40 percent serpentine), and the planar structure is made apparent by alternating laminae of partly serpentinized, relatively coarse, slightly elongate olivine and wider bands of more strongly serpentinized, fine-grained olivine.

Serpentinite also occurs in small isolated bodies in the southern part of the Yreka quadrangle. One that is too small to map is in greenschist near the mouth of a small ravine draining into Moffett Creek at about the corner between secs. 5, 6, 7, and 8, T. 42 N., R. 7 W. Other slightly larger bodies in SE¼ sec. 1, T. 42 N., R. 7 W., and SW¼ sec. 31, T. 43 N., R. 6 W., also crop out along a fault between the Moffett Creek and Gazelle Formations. These isolated serpentinite bodies are enigmatic but can be viewed as supporting evidence for the occurrence of ultramafic rocks beneath the plate of sedimentary rocks, as postulated by Irwin and Lipman (1962) and suggested by the gravity studies of LaFehr (1966).

Most of the gabbro bodies are small and situated on the borders of the ultramafic body. One body is known to occur at the northwest contact of serpentinite with siliceous phyllites of the Stuart Fork Formation, in the south part of sec. 1, T. 44 N., R. 8 W. The other known gabbros are on the amphibolite side of the contact between serpentinite and the belt of amphibolite. Microscopic examination of several specimens shows them to be hornblende gabbros with or without accessory augite. Plagioclase in all specimens is altered, commonly to cloudy gray saussurite or, in some, to clear albite accompanied by clinozoisite.

The Trinity ultramafic body is considered to be part of an Ordovician ophiolite (Irwin, 1973; Hopson and Mattinson, 1973; Lindsley-Griffin, 1973). The serpentinite belt in the Yreka quadrangle is presumably a continuation of the Trinity body and is believed to be Ordovician.

QUARTZ DIORITE AND TRONDHJEMITE

Two small bodies of silicic plutonic rock occur in the southern part of the Yreka quadrangle on the ridge south of Sissel Gulch (secs. 8 and 9, T. 42 N., R. 7 W.) (pl. 1). One of these bodies forms a prominent bluff about 150 m (500 ft) high, locally known as Skookum Butte, that culminates in peak 5382 on the east side of Moffett Creek. Similar rocks crop out in a narrow northeast-trending belt in northwestern China Mountain and northeastern Etna quadrangles.

Megascopically these are light-colored brown-weathering fine-grained, apparently altered quartz-rich granitoid rocks composed of dull-gray feldspar, quartz, splotchy concentrations of dark-greenish-gray hornblende and chlorite and, in some specimens, scattered grayish-green granules of epidote. The rock is commonly cut by many closely spaced fractures and at many places is crumbly and weathers readily to brown gritty soil (grus). Locally the rock is a microbreccia composed of megascopically visible grains of quartz and feldspar in a medium- to dark-gray microscopic matrix.

At least two varieties of the unbrecciated rock, both of which may occur in the same body, are visible under the microscope. At Skookum Butte (SW¼ sec. 8, T. 42 N., R. 7 W.) both a myrmekitic and nonmyrmekitic variety were obtained. The myrmekitic variety (table 4, no. 1) is a fine- to medium-grained rock with a relict hypidiomorphic-granular texture now largely replaced by myrmekite. Subhedral gray turbid albitic plagioclase is enclosed and embayed by interstitial anhedral quartz and myrmekite. The plagioclase is flecked with scaly sericite, and some crystals have cores of epidote, but some of the plagioclase has narrow rims of clear albite. No potassium feldspar was recognized in thin section, but a stained surface was approximately 9 percent very fine interstitial potassium feldspar, which probably occurs in the myrmekitic intergrowths. Epidote also occurs as coarsely crystalline interstitial masses, and clots of chlorite are also accompanied by epidote. Some altered sphene is present, and a few carbonate veinlets cut the rock. The modal composition of the rock is difficult to determine, but it is estimated that quartz is approximately 14 percent, plagioclase 68 percent, potassium feldspar 9 percent, and mafic minerals 8 percent by volume.

The nonmyrmekitic variety (table 4, no. 2) is medium grained and hypidiomorphic granular: the plagioclase tends to be sub-hedral; the other constituents are mostly anhedral. The plagioclase is sodic oligoclase to albite, commonly twinned, and is gray with abundant sericite. Quartz is anhedral, commonly lobate. A trace of potassium feldspar was revealed by selective staining of a sawed surface, but none was recognized in thin section. Mafic minerals include bluish-green hornblende and biotite, both partly replaced by chlorite. Minor amounts of muscovite are present.

TABLE 4.—*Chemical and spectrographic analyses of altered quartz diorite*

[Rapid rock analyses by P. Elmore, J. Kelsey, H. Smith, and J. Glenn. Spectrographic analyses by Chris Heropoulos]

Sample No	1	2
Chemical analyses (weight percent)		
SiO_2	64.2	63.2
Al_2O_3	16.2	14.5
Fe_2O_3	2.1	2.4
FeO	2.5	5.2
MgO	2.0	3.1
CaO	5.3	3.9
Na_2O	3.5	3.4
K_2O	1.5	.76
H_2O^+	2.0	2.6
H_2O^-	.11	.24
TiO_2	.29	.24
P_2O_5	.02	.02
MnO	.10	.16
CO_2	< .05	< .05
Sum	100	100
Specific gravity	2.56	2.75
Semiquantitative spectrographic analyses (parts per million)		
Ba	300	150
Co	10	20
Cr	10	15
Cu	50	100
Ga	20	15
Ni	3	10
Sc	20	50
Sr	300	150
V	100	200
Y	10	15
Zr	70	20
Modes (volume percent)		
Quartz	14	29
Plagioclase	68	48
Potassium feldspar	9	Tr.
Mafic minerals	8	23

Description and location:
1. Altered myrmekitic quartz diorite (Y-39-69). SE¼ sec. 7, T. 42 N., R. 7 W., China Mountain quadrangle.
2. Altered quartz diorite (Y-48-71). SW¼ sec. 8, T. 42 N., R. 7 W., Yreka quadrangle.

Accessory minerals include epidote, sphene, and very minor amounts of opaque minerals, chiefly secondary hematite. The specimen represented by sample 2 in table 4 is cut by a network of fine fractures filled with secondary minerals including prehnite, chlorite, minor sericite, and epidote. The estimated modal composition of the rock from a stained slab is quartz, ·29 percent; plagioclase, 48 percent; and mafic minerals, 23 percent by volume, with a trace of potassium feldspar.

Commonly these rocks have a cataclastic texture. The least broken ones are cut by numerous narrow fractures filled with carbonate and chlorite and, in one specimen, prehnite. Some are crisscrossed by narrow seams of microbreccia, and, even beyond the seams, twinning lamellae of plagioclase are bent and may be microscopically faulted. Quartz is markedly strained. Other specimens from near the borders of the plutons are microbreccias composed of angular fragments of plagioclase and quartz ranging in size from several millimeters to 0.01 mm in a very fine grained brownish-green chloritic matrix accompanied by minor amounts of sericite and traces of epidote.

According to their mineral composition, many of the rocks are altered quartz diotite; however, some that contain less than 10 percent mafic minerals and very sparse or no potassium feldspar are trondhjemitic.

These plutonic bodies occur in mafic and silicic schists and phyllites beneath the Sissel Graywacke and Duzel Phyllite. In the Etna and China Mountain quadrangles they are in fault contact with the Gazelle Formation. They do not intrude the Duzel Phyllite as Wells, Walker, and Merriam (1959) reported. The contact relations of the plutonic rocks with the metamorphic rocks are unknown because of poor exposures. However, the absence of apparent contact metamorphic effects and the brecciated condition of the rock suggest that the plutons are faulted against the schists and phyllites. Possibly they are large clasts in a tectonic melange whose matrix is the mafic and siliceous schists and phyllites.

The same kind of granitic rock occurs at Lovers Leap in the China Mountain quadrangle, where it is found as clasts in a conglomerate of Late Silurian or Early Devonian age (Rohr and Boucot, 1971). Zircon from similar trondhjemitic rocks near Callahan gave concordia intercept ages of 455 to 480 m.y. (Hopson and Mattinson, 1973). Therefore, the granitic rocks in the southern part of the Yreka and nearby in the China Mountain and Etna quadrangles are pre-Silurian, probably Ordovician in age.

METAMORPHIC ROCKS OF ORDOVICIAN(?) AGE

In the vicinity of upper Moffett Creek, metamorphic rocks and limestone occur beneath the Sissel Gulch Graywacke and the Duzel Phyllite in the terrane where the altered quartz diorite plutons are found. Greenschist is the dominant metamorphic rock; there are lesser amounts of siliceous rocks. Some blueschist-facies rocks are also found associated with the greenschist. The same kinds of rocks are found in upper Moffett Creek in the China Mountain quadrangle, and in the Etna quadrangle east and southeast of McConaughy Gulch.

GREENSCHIST

The greenschist is dark greenish gray, fine grained, and well foliated, and in some varieties white feldspathic layers alternate with mafic layers. The microscope shows that the commonest type is albite+(quartz)+sphene+epidote+actinolite+chlorite schist. In some specimens minor amounts of glaucophane accompany the actinolite.

The siliceous metamorphic rocks are composed mainly of quartz and subordinate wispy white mica and chlorite. Typically the quartz is a granoblastic to highly irregular intergrowth with sutured boundaries between grains. Other constituents commonly present in accessory amounts are calcite veinlets and anhedral patches intergrown with quartz, minor sphene, and apatite. Stilpnomelane in minor amounts was identified and some specimens contain prisms of blue sodic amphibole, probably crossite.

BLUESCHIST

Small bodies of blueschist occur in the metamorphic terrane at the southern boundary of the Yreka quadrangle along the upper part of Moffett Creek and in Sissel Gulch. They occur with the greenschist, but their relation with the greenschist is unknown because, although they crop out as small blocks, their contact with surrounding rocks is concealed. Small outcrops occupying less than 10 m^2 occur about 150 m (500 ft) east of Moffett Creek road, 430 m (1,400 ft) south of the section corner common to secs. 5, 6, 7, and 8, T. 42 N., R. 7 W., and in Sissel Gulch about 60 m (200 ft) south of the line between secs. 4 and 9, T. 42 N., R. 7 W. Float blocks of blueschist are found in a gully on the line between secs. 3 and 4, T. 42 N., R. 7 W., below the small area of metamorphic rock in SE¼ sec. 4, T. 42 N., R. 7 W. Specimens are very fine grained, hard, dark-bluish-gray rocks composed of lawsonite and pale-blue amphibole (glaucophane). Numerous other bodies of blue-schist-facies rocks are known in the same metamorphic terrane in the upper part of Moffett Creek (China Mountain quadrangle) and in McConaughy Gulch (Etna quadrangle).

LIMESTONE

Lenticular bodies of limestone ranging from a few meters to as much as 75 to 90 m (250 to 300 ft) wide and 300 m (1,000 ft) long occur with the metamorphic rocks, mainly with the greenschist. Apparently unfossiliferous, the limestone is light gray and rough surfaced, with minor irregular cherty inclusions, and commonly appears to be fragmental.

The terrane of greenschist, siliceous metamorphic rocks, discontinuous bodies of limestone and granitic rock is interpreted as a metamorphosed tectonic melange. This interpretation is supported by the small serpentinite body that occurs, as previously mentioned, in the greenschist in the southwestern part of the quadrangle. Potter, Hotz, and Rohr (1977) called these rocks the schist of Skookum Gulch and assigned them a Silurian(?) and Ordovician(?) age. White mica from a siliceous semischist in this terrane in Etna quadrangle has a potassium-argon age of 431 m.y. (Lower Silurian) (M. A. Lanphere, written commun., 1977). The lawsonite-glaucophane rocks have not been isotopically dated. Perhaps they are coeval with blueschist of Triassic age in the northwest part of the quadrangle, but it seems doubtful that a Triassic metamorphic event would not be reflected in the enclosing schists.

QUARTZ GABBRO AND DIABASE OF UNCERTAIN AGE

Several small dikelike bodies of mafic rock, not shown on plate 1, intrude sedimentary rocks of the Gazelle and Moffett Creek Formations and metamorphosed sedimentary rocks of the Duzel Phyllite and Sissel Gulch Graywacke. These are dark, generally fine-grained rocks that even megascopically have an altered appearance, with dull-gray and somewhat chalky feldspar and rather blotchy mafic minerals.

Under the microscope the rocks have intergranular to subophitic textures, and some of the coarser ones are hypidiomorphic granular. Plagioclase ranges from somewhat hazy albite to gray translucent saussuritized feldspar. Quartz, amounting to 8 or 9 percent in most speciments but as plentiful as 15 to 18 percent in some, is interstitial to and replaces the plagioclase; commonly the quartz and feldspar have a micrographic intergrowth. The principal original mafic mineral is augite, which is a minor relict mineral. Chlorite is plentiful as a replacement of the pyroxene, and some secondary actinolitic amphibole may be present. Sphene, partly altered to leucoxene, is fairly common, and ilmeno-magnetite has been altered to leucoxene pseudomorphs whereas the magnetite is preserved as skeletal gridworks. Prehnite is a common secondary mineral that occurs interstitially and in veinlets; in sample 2, table 5, it replaces albite extensively. Veinlets of calcite are also present.

These dike rocks are classified as gabbros and diabase depending upon their textures. The plagioclase content ranges from 35 to 45 percent, and mafic minerals make up from 45 to 55 percent of the rock. Most contain less than 10 percent quartz and are called quartz-bearing diabase or gabbro; where quartz exceeds 10 percent, they are quartz diabase or quartz gabbro. Some quartz-free diabase also occurs, which may or may not be comagmatic with the quartz-bearing variety. Chemical analyses of gabbro and diabase are presented in table 5.

The age of these dike rocks is not known. They are younger than the lower Paleozoic rocks that they intrude. Possibly they are a manifestation of the Middle and Late Jurassic plutonism that is widespread in adjacent parts of the eastern Klamath Mountains (Lanphere and others, 1968), but the low-grade metamorphism they consistently show suggests that they may be older.

CENTRAL METAMORPHIC BELT

An almost continuous belt of amphibolite crops out east of the serpentinite belt, between the serpentinite and the semischists and sedimentary rocks, for a distance of about 18 km (11 mi) from Yreka to the west edge of the quadrangle. The amphibolite belt continues southward in the Fort Jones and Etna quadrangles for a total distance of about 32 km (20 mi). In the Etna and Fort Jones quadrangles it is accompanied by discontinuous bodies of foliated impure marble and calcareous schists. These rocks are a northward extension of the central metamorphic belt that includes the Salmon and Abrams Formations in the south-central Klamath Mountains.

AMPHIBOLITE

Most of the amphibolite is uniformly greenish gray and fine grained, with a well-developed planar structure and lineation. Locally the rock is medium grained and well foliated with alternating dark mafic and light feldspathic layers, and tight isoclinal shear folds are visible. Mineral species are too fine grained for megascopic identification, although with a hand lens amphibole usually can be recognized.

Microscopically the amphibolite is completely crystalloblastic, finely foliated and lineated, and usually has no visible relict

TABLE 5.—*Chemical analyses of quartz gabbro and diabase*

[Rapid-rock analyses by Lowell Artis. Method used was a single-solution procedure described by Shapiro (1967). Tr., trace]

Sample No.	1	2	3
Chemical analyses (weight percent)			
SiO_2	51.5	55.2	52.6
Al_2O_3	16.9	14.8	14.2
Fe_2O_3	2.2	2.0	2.5
FeO	8.6	9.7	11.5
MgO	4.0	3.6	4.5
CaO	6.3	6.4	7.1
Na_2O	4.8	3.9	2.8
K_2O	.73	.14	.36
H_2O^+	3.2	3.6	3.4
H_2O^-	.42	.19	.31
TiO_2	.89	1.2	1.6
P_2O_5	.20	.17	.19
MnO	.16	.23	.22
CO_2	.11	.04	.03
Sum	100	101	101
Norms (weight percent)			
Q	....	8.5	7.7
Or	4.3	.8	2.1
Ab	40.6	32.6	23.4
An	22.4	22.2	24.8
Di	5.7	6.6	7.2
Hy	11.6	20.0	24.1
Ol	6.2	....	....
Mt	3.2	2.9	3.6
Il	1.7	2.3	3.0
Ap	.47	.40	.44
Cc	.25	.09	.07
Sum	96.42	96.39	96.41
Modes (volume percent)			
Quartz	8	17	....
Albite	44	36	70
Actinolitic hornblende	8	3	Tr.
Clinopyroxene	16	5	8
Sphene	....	3	2
Chlorite	13	19	17
Prehnite	3	16	Tr.
Secondary biotite	....	....	2
Leucoxene	7	....	Tr.
Other alteration products	2	Tr.	Tr.

Locations:
1. Quartz-bearing gabbro (Y-11-73) Cottonwood Creek. Center, sec. 9, T. 43 N., R. 7 W. On contact between Duzel Phyllite and Schulmeyer Gulch sequence.
2. Quartz gabbro (Y-30-71) southeast slope of Bonnet Rock. Dike in sedimentary rocks.
3. Diabase (Y-50-72) roadside in Moffett Creek north of Sissel Gulch, SW¼SW¼ sec. 32, T. 43 N., R. 7 W. Intrudes volcanic wacke of Sissel Gulch Graywacke.

texture. Some specimens, however, have relict phenocrysts of plagioclase, and a few have remnants of a gabbroic texture.

The principal mineral constituents of the amphibolite are pale-green to bluish-green hornblende and colorless to very pale yellowish green epidote, and less commonly clinozoisite. Measurement of refractive indices and optic angles confirms that the amphiboles are hornblende rather than actinolite. Minor amounts of chlorite occur in some specimens. The amount of plagioclase varies considerably between specimans, ranging from a few percent to 30 to 45 percent in some metagabbros. Mostly the plagioclase is clear, anhedral, and untwinned, and its refractive indices show that it is albite. In a few specimens it is so clouded with alteration products its refractive index cannot be measured, but presumably this is also albite. Minor amounts of quartz are also present. Sphene is fairly common as the only fine-grained minor accessory mineral; however, it is not present in every specimen. Opaque metallic minerals are absent except for a few grains in some specimens. Megascopically visible white veinlets of prehnite cut across the foliation in some specimens. The mineral assemblage of these rocks is diagnostic of the albite-epidote-amphibolite metamorphic facies, intermediate between the greenschist and amphibolite facies (Fyfe and Turner, 1966).

The composition of three typical specimens of the amphibolite (table 6) is similar to that of basalt. Metamorphic recrystallization has mostly obliterated original textures and structures so that direct evidence of the nature of the parental rock is not available. However, the texture in some specimens resembles fine-grained augite-hornblende gabbro that occurs at a few places elsewhere in the amphibolite. Probably the amphibolites were derived from submarine basaltic lavas, pyroclastic rocks, and to a somewhat minor extent, gabbro. Potassium-argon age determinations of 390 m.y. and 399 m.y. (Devonian) for the metamorphism were obtained by M. A. Lanphere (written commun., 1976) from hornblende in two amphibolite samples. A similar metamorphic age (380 m.y.) has been established for the Abrams Mica Schist (Lanphere and others, 1968) in the southern part of the Klamath Mountains.

The amphibolite is here correlated with amphibolite in the Abrams Mica Schist on the basis of its appearance, its association outside the Yreka quadrangle with schistose impure marble and calcareous schists, and its isotopic age. Similar rocks were mapped near Etna by Romey (1962) who traced them from Coffee Creek, where they were called the Grouse Ridge Formation by Davis, Holdaway, Lipman, and Romey (1965). The amphibolite might be the correlative of the predominantly amphibolitic Salmon Hornblende Schist, which has the same general distribution in the southern Klamath Mountains and is presumably of the same age, but calcareous rocks are absent from the Salmon.

TABLE 6.—*Chemical and spectrographic analyses of amphibolite*

[Rapid-rock analyses by Paul Elmore; spectrographic analyses by Chris Heropoulos. N, looked for but not found]

Specimen No	1	2	3
Chemical analyses (weight percent)			
SiO_2	47.1	48.4	48.4
Al_2O_3	14.7	16.2	13.9
Fe_2O_3	6.4	2.5	3.0
FeO	6.0	7.0	8.4
MgO	7.1	9.0	8.0
CaO	9.1	9.6	8.9
Na_2O	2.5	3.1	3.3
K_2O	.28	.42	.92
H_2O^+	2.9	1.7	1.7
H_2O^-	.39	.11	.14
TiO_2	2.2	1.0	2.4
P_2O_5	.24	.10	.35
MnO	.16	.16	.16
CO_2	.03	.06	.01
Sum	99	99	100
Specific gravity, bulk	2.98	2.95	3.01
Niggli values			
Al	21	21	19
Fm	50	49	51
C	23	23	22
Alk	6	7	9
Si	113	107	110
K	.07	.07	.15
Mg	.51	.60	.53
Qz	−11	−21	−26
Semiquantitative spectrographic analyses (parts per million)			
Ba	50	50	150
Co	30	30	30
Cr	150	500	200
Cu	150	70	100
Ga	15	15	20
Nb	10	N	15
Ni	100	200	150
Sc	50	50	30
Sr	500	300	500
V	200	150	200
Y	30	20	30
Zr	70	30	100

Description and location:

1. Greenschist (Y-16-72); albite-actinolite-chlorite-epidote-(quartz); central part NE¼ sec. 2, T. 43 N., R. 8 W., Yreka quadrangle.
2. Amphibole (Y-72-72); albite-hornblende-sphene-(prehnite); SE¼ sec. 7, T. 44 N., R. 7 W.
3. Amphibolite (Y-9-72); albite-hornblende-(epidote)-sphene-(prehnite); SE¼ sec. 29, T. 45 N., R. 7 W.

WESTERN PALEOZOIC AND TRIASSIC BELT

The western Paleozoic and Triassic belt is a major lithologic subdivision of the Klamath Mountains province (Irwin, 1960, 1966). In the Yreka quadrangle a greenstone-chert assemblage and the Stuart Fork Formation constitute this subdivision.

GREENSTONE-CHERT ASSEMBLAGE

Rocks here designated the greenstone-chert assemblage occur north and west of the serpentinite belt. This assemblage is composed predominantly of metavolcanic rocks with interbedded sedimentary rocks, including chert, shale, and argillite, and minor limestone.

METAVOLCANIC ROCKS

In this area the metavolcanic rocks or greenstones are commonly volcanic breccias, but massive units, presumably flows, are plentiful, and locally some exposures show pillow structure. In hand specimens the rocks are greenish gray, fine grained to microscopic, locally porphyritic, and some, especially the volcanic breccias, are notably amygdaloidal or, on weathered surfaces, vesicular. Except for small dark-greenish-gray phenocrysts of pyroxene, amphibole, or chlorite, and calcite amygdules, the minerals and textures of the rocks are visible only under the microscope. In thin sections, however, the textures and minerals typical of greenstone or spilite are apparent.

Plagioclase in the metavolcanic rocks commonly is in tiny, randomly oriented laths in the groundmass; it seldom occurs as phenocrysts. Commonly it is hazy and crowded with submicroscopic alteration products, yet in some specimens the laths are relatively clear and may exhibit lamellar twinning. Refractive index measurements show that the plagioclase is highly sodic, usually approaching pure albite, but probably a sodic oligoclase in some specimens.

Some of the greenstones contain unaltered pyroxene, apparently augite, occurring between the plagioclase laths and also as small phenocrysts or microphenocrysts. Minor chlorite may also be present, as well as accessory clinozoisite or epidote. Albite, pyroxene, and chlorite are the diagnostic mineral assemblage of spilite (Yoder, 1967). In other specimens the original pyroxene is replaced by pale-green actinolitic amphibole. Quartz, prehnite, sphene partly altered to leucoxene, apatite, black metallic minerals—presumably ilmenite-magnetite—and pyrite may occur in variable amounts as minor accessory minerals. Calcite is commonly seen filling fractures and amygdules.

TABLE 7.—*Chemical and semiquantitative spectrographic analyses of greenstone*

[Rapid-rock analyses, samples 2 and 3, by Paul Elmore, Sam Botts, Lowell Artis; sample 1, by Paul Elmore, Sam Botts, Lowell Artis, G. W. Chloe, J. L. Glenn, James Kelsey. Spectrographic analyses, samples 2 and 3, by W. B. Crandell; sample 1 by Chris Heropoulos]

Sample No (see p. 47)	1	2	3
Chemical analyses (weight percent)			
SiO_2	45.7	48.1	50.6
Al_2O_3	13.0	15.0	15.0
Fe_2O_3	5.0	1.2	.66
FeO	6.4	9.0	7.4
MgO	12.2	5.7	9.5
CaO	7.2	9.3	7.4
Na_2O	2.3	4.9	3.8
K_2O	1.1	1.4	.91
H_2O^+	3.8	1.5	3.4
H_2O^-	.50	.08	.20
TiO_2	2.2	2.6	.87
P_2O_5	.35	.39	.11
MnO	.18	.14	.15
CO_2	.08	.85	.14
Sum	100	100	100
Specific gravity, bulk	2.99	2.98	2.95
Semiquantitative analyses (parts per million)			
Ba	200	300	30
Ce	100	0	0
Co	70	70	50
Cr	700	500	500
Cu	70	150	150
Ga	15	100	100
La	30	0	0
Nb	30	10	0
Ni	300	100	150
Pb	0	10	70
Sc	30	30	50
Sn	0	5	15
Sr	300	500	200
V	150	300	200
Y	2	2	1.5
Zr	150	100	30
Norms (weight percent)			
Or	6.5	8.3	5.4
Ab	19.6	27.0	32.2
An	22.0	14.8	21.2
Ne	....	7.8	....
Di	8.7	19.2	11.3
Hy	17.3	....	5.7
Ol	9.5	11.9	17.7
Mt	7.3	1.7	.96
Il	4.2	4.9	1.7
Ap	.83	.92	.26
Cc	.18	1.9	.32
Sum	96.11	98.4	96.74

The metavolcanic rocks are a continuation of similar rocks that are exposed in the Condrey Mountain quadrangle (Hotz, 1967), the Hornbrook quadrangle, and in the Fort Jones quadrangle. Analyses of two specimens from the Hornbrook quadrangle and one from the Condrey Mountain quadrangle are presented in table 7. These rocks are classified as alkalic basalts because of their relatively high content of Na_2O+K_2O relative to SiO_2 (Kuno, 1968). They also contain considerably more K_2O than is usually found in spilites or in oceanic tholeiites. It is uncertain how much chemical change has taken place in the rocks since they were erupted. They may have undergone deuteric alteration as they cooled in the presence of seawater, although Moore (1965) has noted that there were no great changes in submarine lavas in Hawaii. Coleman and Lee (1963) conclude that there is little or no change in bulk chemistry in basalts metamorphosed to blueschists, and Bailey, Irwin, and Jones (1964) believe that the composition of massive Franciscan greenstone is close to that of the original magma.

SEDIMENTARY ROCKS

All the sedimentary rocks associated with the metavolcanic rocks are fine grained and are mainly chert, shale, and argillite. Some siltstone and, rarely, very fine grained sandstone occur locally. Limestone is the least abundant.

Chert occurs in some fairly persistent units several hundred meters thick that can be followed along strike for thousands of meters. Most commonly the chert is rhythmically bedded with layers 5 cm to 10 cm (2 to 4 in.) thick separated by interbeds of greenish phyllite 5 to 15 mm (0.2 to 0.6 in.) thick. Some chert bodies are markedly lenticular with thicknesses of several tens of meters and lengths of a few hundred meters. The more lenticular masses tend to be thick bedded to massive, although lenses of rhythmically bedded chert also occur. The chert ranges from white to dark gray to black; green and red chert are not common. Fine color banding is common in individual chert layers. Because of metamorphism, the chert is no longer cryptocrystalline and has lost its subvitreous luster and typically conchoidal fracture. It is now a very fine grained to microcrystalline quartz intergrowth that gives the rock a granular appearance and an angular, blocky fracture.

Shale and argillite occur with chert and are interbedded also

Sample description and location:
1. Fragmental greenstone (HO–20–66); albite-augite-actinolite-chlorite-(epidote); 0.3 km (0.2 mi) south of Paradise Craggy Lookout, Hornbrook quadrangle.
2. Porphyritic metavolcanic rock (CM-16-61); albite-actinolite-pale biotite-calcite-sphene; NW¼ sec. 5, T. 45 N., R. 8 W., Condrey Mountain quadrangle.
3. Porphyritic greenstone (CM-2-64); albite-augite-chlorite; SW¼ sec. 18, T. 46 N., R. 6 W., Hornbrook quadrangle.

with the metavolcanic rocks. Much of the shale is light gray, siliceous, and commonly somewhat phyllitic. It commonly has a well-developed slaty cleavage, and at some places a well-developed pencil cleavage owing to intersecting cleavages. Dark-gray to black argillite is also interbedded with the volcanic rocks in places. Locally, rather distinctive reddish-brown shale and siltstone are interbedded with the volcanic rocks.

The limestone bodies are generally small and distinctly lenticular, although some are in belts where several discontinuous bodies occur more or less on strike. The bodies that are interbedded with the shale, phyllite, or chert are 3 to 6 m (10 to 20 ft) thick and a few tens of meters to as much as 300 m (1,000 ft) long. They are commonly crystalline and weather light gray with a characteristic "licked salt" appearance. On fresh fractures they are dark gray to black and mottled black and white. Some are graphitic. Small pods and streaks of limestone also occur with the volcanic rocks. In places, limestone is intermixed with the fragmental volcanic rocks, and it also occurs at the blocky or highly vesicular interface between flows. The limestones are commonly dense, fine grained, and impure. Some are oolitic.

AGE

The greenstone-chert assemblage in Yreka quadrangle has long been regarded as part of the Applegate Group of southwestern Oregon, of Triassic age. Undoubtedly it is at least in part continuous with the Applegate, and is part of the western Paleozoic and Triassic belt. Recently its age has been more firmly established by the discovery of Late Permian fusulinids in limestone in the Hornbrook quadrangle about 15 km (9 mi) northeast of Yreka (Elliott and Bostwick, 1973). The limestone is a small lentil contained in a sequence of metachert, greenstone, quartzite, and metaconglomerate that is continuous with the metavolcanic and sedimentary rocks in the northwest part of the Yreka quadrangle.

STUART FORK FORMATION

DISTRIBUTION

The Stuart Fork Formation was named by Davis and Lipman (1962) for the excellent exposures along the Stuart Fork of Trinity River between Deep and Van Matre Creeks in the Trinity Lake quadrangle. In the Yreka-Fort Jones area it forms a continuous belt between serpentinite and the greenstone-chert sequence, trending in a northeast-southwest direction for about 50 km (31 mi) from the southeast part of the Hornbrook quadrangle where it emerges from a Tertiary volcanic cover, and across the north-western part of the Yreka quadrangle and central part of the Fort Jones quadrangle to its termination along a northwest-trending fault against rocks of the western Paleozoic and Triassic belt in the western part of the Fort Jones quadrangle. The Stuart Fork Formation ranges in width from approximately 400 m (0.25 mi) to about 5 km (3 mi). It is thrust over the greenstone-chert assemblage on the Soap Creek Ridge thrust fault.

LITHOLOGY

In the Yreka-Fort Jones area the Stuart Fork Formation is composed predominantly of phyllitic quartzite and lesser amounts of blueschist. The blueschist occurs as layers within the phyllitic quartzite and as isolated pods of metabasalt surrounded by the siliceous rocks (Hotz, 1973). A few small lenses and pods of marble occur with the phyllitic quartzite.

The phyllitic quartzites are thinly layered rocks composed of alternating layers of quartzite and micaceous folia (fig. 6). The quartz laminae are from 1 mm to as much as 2.5 mm (0.04-1 in.) thick. The intervening micaceous laminae are seldom more than 1 mm wide, except where thickened on the crests of folds, and may be little more than films a fraction of a millimeter thick. Folding is conspicuous in nearly every exposure and ranges from rather open, wavy, low-amplitude corrugations to tight isoclinal folds, and at many places to highly complex discontinuous and disruptive plications.

Under the microscope it is apparent that the principal mineral of these rocks is quartz in a fine-grained (0.1 mm average) crystallo-blastic mosaic. Muscovite is the most common micaceous mineral, but subordinate amounts of pale chlorite and less commonly minor stilpnomelane may be present. The micaceous minerals are seldom more abundant than 10-15 percent and never more than 25 percent. Some specimens contain a small amount of albite and some have trace amounts of lawsonite.

A less common but rather distinctive variety of phyllitic quartz-ite has a pale-reddish-brown sheen on micaceous cleavage surfaces from microscopic hematite intimately associated with muscovite. Some dark quartz-mica phyllites are graphitic.

The phyllitic quartzites are interpreted as metamorphosed rhythmically layered cherts because of their thin primary layering and high modal quartz.

The most conspicuous bodies of blueschist are metabasalts that crop out as bold, isolated bodies surrounded by phyllitic quartzite. The bodies range in size from 3 m (10 ft) in many exposures to 300 m

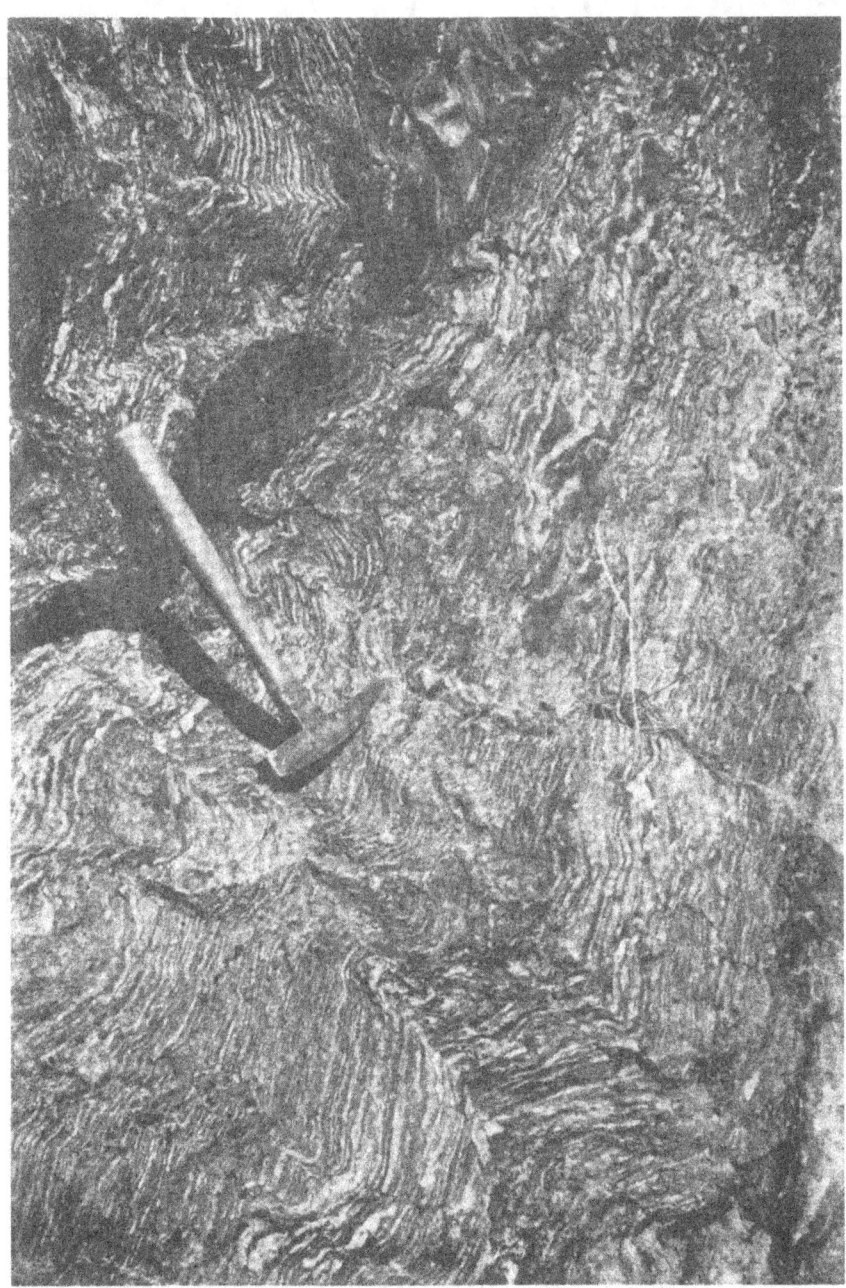

FIGURE 6.—Phyllitic quartzite of Stuart Fork Formation.

(1000 ft) long and more than 100 m (330 ft) wide. The metabasalts commonly are very fine grained, hard, and massive, although some are phyllitic or schistose. Commonly, original structures and textures are obliterated by recrystallization, but in some thin sections faint relics of an ophitic or diabasic texture are visible. The specific gravity of these rocks ranges from 2.91 to 3.14; the average of 19 measurements is 3.08.

Glaucophane is plentiful and is predominant in some of the metabasalt; lawsonite is present in all specimens as strongly euhedral crystals; sphene is also ubiquitous. Minor quantities of white mica and albite are present. Jadeitic pyroxene is seen in a few thin sections, but it is not a common constituent; some metabasalts contain a green pyroxene identified as aegirine. Pumpellyite is found in only a few thin sections. Calcium carbonate is not abundant, but small amounts occur in some specimens. X-ray data indicate that the carbonate is calcite. Pyrite is a common minor constituent, although there is seldom more than one percent. Magnetite, ilmenite, and hematite were not found.

Siliceous metasedimentary rocks containing lawsonite and a sodic amphibole are less conspicuous than the metabasalts because they are commonly interlayered with phyllitic quartzite. Probably they originally were impure quartzites or graywackes, for the composition of relatively pure quartzite or chert is not suitable for the formation of lawsonite. The specific gravity of eight samples averages 2.71.

In hand specimen the rocks are fine grained and foliated with thin alternating dark and light laminae parallel to the cleavage. Under the microscope the crystalloblastic texture is clear. Anhedral interlocking quartz with irregular to sutured boundaries predominates. In some specimens the quartz grains are equidimensional, but in others they are somewhat elongate with nearly parallel arrangement, and idioblastic lawsonite is concentrated in thin laminae. In some specimens the laminae are fairly continuous; in others they are discontinuous wispy streaks and lenses. The amphibole needles commonly wrap around and enclose the lawsonite crystals. In some rocks the laminae are highly contorted by microfolds. Crossite is the common sodium amphibole in the metasedimentary rocks, but in some, presumably of lower original iron content, pale-blue glaucophane was formed instead. White mica, although common, is a minor constituent, but in some specimens it is more plentiful than the sodium amphibole. Traces of chlorite and stilpnomelane are common, and a chlorite-lawsonite-quartz phyllite without blue amphibole occurs locally. Small grains of garnet occur in a few specimens.

AGE AND CORRELATION

There is no direct stratigraphic evidence of the age of the Stuart Fork Formation, and the sparse limestone bodies within it have yielded no fossils. In the Yreka-Fort Jones area the formation occupies a thrust plate overlying the greenstone-chert sequence of Permian age and is beneath a tabular serpentinite body of Ordovician(?) age. In the southern Klamath Mountains it underlies metamorphic rocks of the Salmon and Abrams Formations, which override it on a regional thrust fault. In the Scott Bar Mountains in the Fort Jones and Condrey Mountain quadrangles, klippen of the Stuart Fork Formation are apparently intruded by Middle or Late Jurassic (145 to 155 m.y.) granitic plutons, but to the south in the Trinity Alps and Salmon Mountains the formation is intruded by granitic plutons 127 to 140 m.y. old (Lanphere and others, 1968).

A Middle to Late Jurassic age has been postulated for the metamorphism of the Stuart Fork Formation, on the basis of some potassium-argon age determinations ranging from 133 to 157 m.y. for samples collected in the southern Klamath Mountains (Lanphere and others, 1968). In the Yreka-Fort Jones area, however, potassium-argon measurements by M. A. Lanphere on mica from crossite-lawsonite-quartz-white mica (phengite?) schist yielded ages of approximately 220 m.y. (Middle Triassic) (Hotz and others, 1977).

On the basis of lithologic, metamorphic, and structural similarities, the Stuart Fork has been correlated with the western Paleozoic and Triassic belt in the southern Klamath Mountains (Davis and Lipman, 1962; Davis and others, 1965; Davis, 1968). In the Yreka-Fort Jones area where the Stuart Fork Formation is predominantly composed of phyllitic quartzite, no direct correlation can be made with the Permian greenstone-chert sequence in which the predominant rocks are metavolcanic. Chert and siliceous shale occur within the Permian sequence, however, and are plentiful in the "upper" part near the thrust fault beneath the Stuart Fork Formation where, in places, they are considerably deformed and phyllitic and closely resemble some of the siliceous phyllites in the Stuart Fork Formation. The widespread distribution and persistence of predominantly siliceous sedimentary rocks, however, indicate that the Stuart Fork is unlike other units in the western Paleozoic and Triassic belt (Irwin, 1972). Conceivably, it is a facies not represented in the Paleozoic and Triassic belt as now exposed but is an allochthonous unit carried in on the thrust plate.

CRETACEOUS ROCKS

HORNBROOK FORMATION

The Hornbrook Formation was defined by Peck, Imlay, and Popenoe (1956) as Upper Cretaceous marine sedimentary rocks exposed along the valley of Cottonwood Creek near Hornbrook, Calif., a town about 21 km (13 mi) north of Yreka. These rocks are correlative with the Chico Formation in the Sacramento Valley (Williams, 1949). They were studied at the type locality by Jones (1959) and Elliott (1971). No detailed studies have been made of the Hornbrook Formation in the Yreka quadrangle, where it occurs as discontinuous remnants.

In the Yreka quadrangle rocks correlated with the Hornbrook Formation crop out in the low hills east, northeast, and southeast of Yreka, where they unconformably overlie serpentinite and the Schulmeyer Gulch sequence. The strata dip gently eastward and are overlapped by alluvium in Shasta Valley. A small erosional remnant southwest of Yreka on Soap Creek Ridge rests unconformably on serpentinite and the Stuart Fork Formation.

Typically, the Hornbrook Formation in the Yreka area is composed of greenish-gray, buff- to brown-weathering, firmly cemented coarse- to medium-grained sandstone. Massive beds a few meters thick alternate with layers that range from 1 or 2 cm (0.4 or 0.8 in.) to several centimeters in thickness. The beds tend to be conglomeratic near the base with clasts that range from pebbles to boulders, and conglomerate layers occur elsewhere in the section. The conglomerate clasts closely reflect the lithology of the nearby subjacent rocks and therefore include abundant quartz, quartzite and chert, phyllitic quartzite, and some metavolcanic rocks and serpentinite. Fine-grained sandstone and dark-gray to black mudstone and shale become plentiful in the upper part of the exposed section.

At the type locality of the Hornbrook Formation a thickness of approximately 760 m (2,500 ft) was measured by Peck, Imlay, and Popenoe (1956), but Jones' (1959) and Elliott's (1971) studies in this same area showed that beds that Peck, Imlay, and Popenoe (1956) regarded as Eocene are actually part of the Cretaceous sequence. Jones (1959) estimated a total thickness of 1,460 to 1,490 m (4,800 to 4,900 ft); Elliott (1971) measured approximately 1,220 m (4,000 ft) in the same area. The Hornbrook Formation is relatively thin where it is exposed in the Yreka quadrangle, owing partly to erosion but probably mainly because the section was originally thinner here. Only the lower part of the section exposed in the type locality is represented in the southern exposures. East of Yreka the Hornbrook Formation is approximately 250 m (800 ft) thick; the remnant on Soap Creek Ridge is only about 40 m (130 ft) thick.

Paleontologic evidence indicates that the Hornbrook Formation in its type locality is Late Cretaceous. Peck, Imlay, and Popenoe (1956) reported fossils of Cenomanian, Turonian, and Campanian age. According to Jones (1959) and Elliott (1971) the lower beds contain upper Turonian and lower Coniacian fossils. Above an unconformity, beds previously mapped as the Eocene Umpqua Formation are of middle Campanian to Maestrichtian(?) age, according to Elliott (1971). The strata in the Yreka area have not been correlated directly with the type locality, but they resemble the coarse clastic rocks in the lower part of the section and therefore may be upper Turonian and lower Coniacian.

TERTIARY ROCKS

Rocks of Tertiary age crop out in the eastern part of the Yreka quadrangle. They include volcanic rocks and remnant patches of conglomerate assigned a Tertiary(?) age.

VOLCANIC ROCKS

The volcanic rocks in the eastern part of the Yreka quadrangle are apparently a continuation of the Western Cascade volcanic

FIGURE 7.—Volcanic hills and alluvial flats in Shasta Valley. A, Hill of tuff breccia
conical peak of Gregory

series mapped by Williams (1949) in the Macdoel 30-minute quadrangle. In Oregon the Western Cascade volcanic series ranges in age from late Eocene to late Miocene. Near Hornbrook, north of Yreka, plant remains of late Eocene or Oligocene age occur in tuff forming the lowermost 12 m (40 ft) of the volcanic series (Williams, 1949). Volcanic rocks in the Yreka quadrangle include an extensive field of tuff breccia, some flows, and a small intrusive body. Tuff breccia and flows are not mapped separately (fig. 1), but an intrusive body is distinguished in the northeast part of the quadrangle.

TUFF BRECCIA

The many small conical hills in Shasta Valley (fig. 7), on the east edge of the quadrangle, are erosional remnants of an extensive blanket of andesitic tuff breccia. The thickness of the breccia is unknown, but to the east at Sheep Rock in the Macdoel quadrangle the breccias are about 490 m (1,600 ft) thick (Williams, 1949).

Surfaces underlain by tuff breccia are strewn with blocks of volcanic rock and resemble weathered and eroded volcanic flows. In fresh exposures such as roadcuts and quarries, however, it is apparent that flows are exceedingly rare and in most places absent. Instead, the exposures reveal a jumbled accumulation of blocks of volcanic rock as much as 1 to 1.5 m (3 to 5 ft) across in an

surrounded by alluvium. *B*, Hills of tuff breccia and flows, alluvial flats, and Mountain plug on the left.

unconsolidated sandy to clinkery tuffaceous matrix. The blocks are composed of dark-gray, gray, and grayish-red massive to vesicular or scoriaceous porphyritic volcanic rock. At some places the blocks are closely packed, with little tuffaceous interstitial material; at other places larger blocks are distributed in a fine tuffaceous matrix; and some excavations expose roughly stratified crystal-vitric tuff containing pebbles, cobbles, boulders, and subrounded blocks of volcanic rock. Stratification in the tuff breccia is not generally apparent, but in places the alternation of layers of slightly different grain size makes a crude bedding recognizable. Flow units apparently occur within the tuff breccia at a few places; they are very uncommon and are not exposed well enough to show their thickness or extent, but they seem to be thin and very limited.

The lava blocks are porphyritic andesites containing mainly plagioclase and pyroxene phenocrysts. The fine-grained groundmass is composed of very small to microscopic laths of plagioclase, minor euhedral pyroxene crystals, and scattered magnetite grains in a gray to brownish glass that contains hazy plagioclase microlites and is commonly heavily charged with magnetite dust. Most commonly the plagioclase phenocrysts are strongly zoned labradorite ranging from about An_{58} to An_{65}; in a given specimen the groundmass feldspar is slightly more sodic than the phenocrysts, but is still labradorite. Hypersthene and augite occur together, but hypersthene is more plentiful than the clinopyroxene. Chemical analyses of these rocks confirm their petrographic classification as andesites (table 8).

FLOWS

East and south of Montague, excepting Gregory Mountain, the terrane is underlain predominantly by gray andesitic flow rocks. These are nonvesicular porphyritic rocks composed predominantly of plagioclase and minor pyroxene phenocrysts in a microcrystalline groundmass (table 8, nos. 5, 6). The plagioclase phenocrysts are strongly zoned in normal or oscillatory fashion. They range in composition from An_{50} to An_{60} (labradorite); groundmass crystals are also labradorite, about An_{56}. Clinopyroxene (augite) is subordinate to orthopyroxene (hypersthene); both occur in minor quantities in the groundmass as well as in phenocrysts.

A small flow remnant caps the ridge northwest of Bonnet Rock where it is situated about 365 m (1,200 ft) above the volcanic rocks of Shasta Valley near Gazelle. The rock is a medium-gray, sparsely porphyritic basaltic andesite. The phenocrysts are olivine, set in an intergranular matrix of plagioclase laths and grains of augite and minor hypersthene. The plagioclase is andesine-labradorite, approximately An_{50}.

PLUG AT GREGORY MOUNTAIN

A conical peak called Gregory Mountain (fig. 7B) about 2 km (1 mi) southeast of Montague is the outcrop of a cylindrical body of very uniform fine-grained porphyritic hornblende andesite. The rock has been slightly altered deuterically so that megascopically

TABLE 8.—*Chemical analyses of Tertiary volcanic rocks, Yreka quadrangle*

[Rapid-rock analyses by Lowell Artis. Method used was a single-solution procedure described by Shapiro (1967)]

Sample No	1	2	3	4	5	6	7	8
Chemical analyses (weight percent)								
SiO_2	61.0	61.0	61.7	61.7	61.7	62.2	62.8	65.5
Al_2O_3	17.4	17.5	17.4	17.5	17.6	17.5	17.5	16.4
Fe_2O_3	1.6	1.9	2.5	3.5	1.9	2.3	2.9	2.5
FeO	2.9	2.7	2.4	.46	2.4	2.2	1.7	1.2
MgO	3.3	3.1	3.4	2.4	3.0	3.1	3.0	1.2
CaO	6.8	6.9	6.5	6.6	6.7	6.5	6.7	3.7
Na_2O	4.3	4.3	4.2	4.5	4.3	4.3	4.2	4.5
K_2O	.80	.95	.90	.80	.90	1.1	.77	1.7
H_2O^+	.80	.29	.34	.24	.46	.31	.25	1.3
H_2O^-	.03	.10	.02	.03	.12	.02	.00	1.0
TiO_2	.65	.68	.70	.67	.66	.70	.70	.38
P_2O_5	.21	.24	.24	.28	.24	.23	.24	.21
MnO	.06	.06	.06	.05	.06	.06	.08	.06
CO_2	.01	.01	.01	.01	.01	.01	.01	.04
Sum	100	100	100	99	100	101	101	100
Norms (weight percent)								
Q	14.1	14.1	15.9	17.0	15.5	15.7	18.2	23.9
Or	4.7	5.6	5.3	4.8	5.3	6.5	4.5	10.1
Ab	36.4	36.5	35.4	38.6	36.4	36.2	35.2	38.2
An	25.9	25.7	25.9	25.5	26.1	25.1	26.4	16.8
Di	5.2	5.6	3.7	4.2	4.4	4.3	3.8	....
Hy	8.8	7.5	8.0	4.1	7.2	6.7	5.6	3.0
Mt	2.3	2.8	3.6	....	2.8	3.3	3.7	3.0
Hm	....	....	....	....	....	....	....	.46
Il	1.2	1.3	1.3	1.1	1.3	1.3	1.3	.72
Ap	.50	.57	.57	.67	.57	.54	.56	.50
Cc	.02	.02	.02	.02	.02	.02	.02	.09

Location of samples:

1. (Y-66-72) Dark-gray breccia block in quarry, NW¼ sec. 27, T. 45 N., R. 6 W.
2. (Y-73-72) Gray, porphyritic flow unit in quarry; line between sec. 34, T. 45 N., R. 6 W., and sec. 3, T. 44 N., R. 6 W.
3. (Y-63-72) Roadcut, NE¼ sec. 33, T. 45 N., R. 6 W.
4. (Y-84-72) Reddish-brown, highly vesicular andesite from breccia block; bulldozer cut, SE¼ sec. 15, T. 44 N., R. 6 W.
5. (Y-83-72) Platy porphyritic flow unit; NE¼ sec. 14, T. 44 N., R. 6 W.
6. (Y-64-72) Dikelike fine-grained andesite in quarry, NW¼ sec. 27, T. 45 N., R. 6 W.
7. (Y-65-72) Fine-grained gray breccia block in quarry, NW¼ sec. 27, T. 45 N., R. 6 W.
8. (Y-2-73) Gray, altered andesite from top of Gregory Mountain (SW¼SW¼ sec. 26, T. 45 N., R. 6 W.

it has small chalky feldspar phenocrysts and hornblende phenocrysts with dull reddish-brown rims and fresh-appearing cores. Under the microscope partly altered labradorite (An_{68}) phenocrysts, some with narrow rims of more sodic plagioclase, and brown hornblende with altered opaque magnetitic rims are in a microcrystalline groundmass of stubby feldspar (apparently mostly plagioclase), very minor rounded quartz grains, and an unidentified green micaceous mineral, possibly a clay. Not seen in thin section but visible in hand specimen are very small rounded grains of pink garnet that are probably xenocrysts; they constitute less than 0.5 percent of the rock. An analysis of this rock (no. 8, table 8) shows it to be slightly more silicic than the andesite tuff breccias and flows. Its altered condition is reflected in its high water content relative to unaltered andesites of the tuff breccias.

CONGLOMERATE

Some small remnants of a sedimentary unit, predominantly conglomerate, occur in the eastern part of the Yreka quadrangle. Some outcrops are found on the edge of Shasta Valley about 5 km (3 mi) south of Grenada, and two more cap ridges east and west of the upper part of Cram Gulch about 8 km (5 mi) southwest of Grenada. A third occurrence is on the edge of the valley about 4 km (2.5 mi) east of Yreka. The largest deposit is near the head of Cram Gulch in NW¼ sec. 7, T. 43 N., R. 6 W., at the site of Pythian Cave, a shallow natural cavern.

These sedimentary remnants are firmly cemented conglomerate with minor coarse sand lenses. They are nearly horizontal; where bedding can be recognized, it dips very gently and could well be an original depositional feature. Some crossbedding is recognizable. The conglomerate clasts are well rounded and range in size from pebbles to boulders. They include predominantly Tertiary volcanic rock, in addition to lesser amounts of chert, quartzite, serpentinite, and sandstone. In places, for example the locality 5 km south of Grenada, the clasts seem to be exclusively of volcanic origin. Mostly they are fresh porphyritic nonvesicular rocks, apparently of basaltic and andesitic composition. Some varieties have become soft through weathering. None appears to be vesicular or or scoriaceous as are the fragments in the tuff breccia.

These deposits are as much as about 425 m (1,400 ft) above the floor of Shasta Valley and have a maximum thickness of perhaps 120 m (400 ft), commonly considerably less. The deposit at Pythian Cave was originally correlated (Mack, 1960) with the Cretaceous Hornbrook Formation (previously Chico), and identical beds 5 km (3 mi) south of Grenada were assigned to the Quaternary System

(Mack, 1960). The beds cannot be as old as Cretaceous because they contain Tertiary volcanic rocks of the Western Cascade Province. They are more firmly cemented than Quaternary deposits in Shasta Valley and are assigned a questionable Tertiary age.

QUATERNARY DEPOSITS

Two kinds of Quaternary deposits are delineated in the Yreka-Fort Jones area: a distinctive alluvial deposit in Shasta Valley, and alluvium and colluvium elsewhere along streams, in valley bottoms, and on lower slopes bordering the valleys.

ALLUVIUM OF SHASTA VALLEY

In Shasta Valley an alluvial deposit extends from south of Gazelle (possibly as far south as west of Edgewood, 5 km (3 mi) northwest of Weed) to somewhere north of Montague. From near Gazelle to about 4 km (2.5 mi) north of Grenada it forms a smooth surface that slopes very gently (5.6 m/km, 30 ft/mi) northward and through which the little conical hills and mounds of tuff breccia protrude. It is barely incised by the small meandering stream of Willow Creek, but about 4 km (2.5 mi) north of Grenada the Shasta River and, a little farther north, the Little Shasta River have eroded a channel in this older alluvium. The channel deepens northward until near the edge of the Yreka quadrangle the relief between the Shasta River flood plain and the older surface is more than 12 m (40 ft).

The alluvium is exposed at several places in roadcuts and in shallow bulldozed trenches into which groundwater flows and is stored for irrigation. The deposit is characteristically a compact but uncemented silt to very fine to fine grained sand, poorly stratified to nonstratified, and containing scattered waterworn pebbles, cobbles, and less commonly boulders. The clasts are of two general kinds: (1) those from the Klamath Mountains province, which include serpentinite, gabbro, phyllite, chert, and quartzite and (2) clasts of volcanic rock from the Cascade province.

The deposit is restricted to Shasta Valley and does not extend into the valleys that drain the hills to the west. It is possibly of glaciofluvial origin, laid down by meltwater streams from glaciated areas in the high parts of the Klamath Mountains known as The Eddys, southwest of Weed, and from the slopes of Mount Shasta.

OTHER ALLUVIAL DEPOSITS

Unconsolidated alluvial gravel occupies the larger stream beds and some of the smaller ones. Colluvium, or slope wash, mantles the slopes in most places, but the bedrock can usually be determined from occasional exposures and the nature of the float. On the lower slopes of some wider valleys, however, colluvium effectively conceals bedrock, forms broad aprons, and fills valley bottoms. Although most of these alluvial deposits have been prospected for gold, only those streams that drain areas underlain by rocks of the western Paleozoic and Triassic belt yielded any placer deposits. None occur in the area south of Greenhorn Creek and east of Scott Valley. In the Yreka quadrangle Greenhorn Creek was auriferous, and placers in Yreka Creek and the gulches west of it were extensively mined.

STRUCTURAL GEOLOGY

GENERAL FEATURES

The Paleozoic sedimentary rocks of the eastern Klamath belt in the Yreka quadrangle overlie what is interpreted to be a tabular body of ultramafic rocks whose exposed upturned edge is represented by the northeast-trending belt of serpentinite. The melange of greenschist, marble lenses, minor siliceous metamorphic rocks, and the small bodies of altered quartz diorite or trondhjemite in the southern part of the Yreka quadrangle is part of the basement.

Supporting evidence for the presence of ultramafic rocks beneath the Paleozoic rocks in the Yreka area is found in the results of gravity studies by LaFehr (1966). His calculations support the interpretation that a sheetlike ultramafic body whose upper surface is no deeper than about 4 km (2.5 mi) is responsible for the gravity anomaly. Narrow, discontinuous, highly sheared bodies of serpentinite along the northeast-trending fault between the Moffett Creek Formation and the Gazelle Formation west of Bonnet Rock and another isolated body of serpentinite, too small to map, in greenschist along the upper part of Moffett Creek may also support this interpretation.

The concept that a tabular ultramafic body underlies rocks of the eastern Klamath belt in this region was first postulated by Hershey (1901), who stated that "***in the Scott Valley region*** a sheet of serpentinite appears to underlie the Devono-Carboniferous over at least several square miles." Irwin and Lipman (1962) pictured a continuous subhorizontal sheet of ultramafic rocks separating rocks of the central metamorphic belt from structurally overlying rocks of the western Paleozoic belt. They interpreted the northeast-trending belt of serpentinite in the Yreka-Fort Jones area as the upturned edge of a great sheet of ultramafic rocks that underlies the plate of Paleozoic rocks and, surfacing again to the southeast, constitutes the great mass of ultramafic rock called the Trinity ultramafic body, now recognized as part of an Ordovician ophiolite complex (Irwin, 1973; Hopson and Mattinson, 1973; Lindsley-Griffin, 1973).

Northwest of the northeast-trending serpentinite belt the intensely deformed siliceous phyllite and blueschist of the Stuart Fork Formation is thrust northwestward over the Permian greenstone-chert assemblage. On the east, the northeast-trending Paleozoic rocks of the eastern Klamath belt are overlapped by the Tertiary volcanic rocks of the Cascade province.

STRUCTURES OF THE EASTERN KLAMATH BELT

The Paleozoic rocks east of the serpentinite belt occur in several structural units of contrasting deformational style and metamorphic grade that have been juxtaposed by horizontal displacement along thrust faults. Closely folded weakly metamorphosed phyllites of the Duzel Phyllite and Sissel Gulch Graywacke are thrust over unmetamorphosed rocks of the Gazelle and Moffett Creek Formations in the southeastern part of the quadrangle. The Moffett Creek Formation in turn is in a higher thrust slice superposed on the Duzel in the western part of the area. The highest plate of all is represented by the limestone of Duzel Rock, which occupies klippen that overrode the thrust slice of the Moffett Creek Formation.

The relation of the Paleozoic sedimentary rocks to the eastward-dipping amphibolite belt is somewhat ambiguous because if the amphibolite is part of the central metamorphic belt it is not in normal structural sequence; however, the sedimentary rocks are interpreted to be faulted against the amphibolite and its juxtaposition may be the result of complex thrust faulting. At places to the west in the Fort Jones and Etna quadrangles, quartzite and chert beds are cut out against the amphibolite along a steep east-dipping contact. At a few places slivers of sheared serpentinite occur at the contact or within the amphibolite. Further evidence for a tectonic relation is the difference in metamorphic grade: the sedimentary rocks are protomylonites to semischists of the lower greenschist facies, in contrast to the higher grade albite-epidote amphibolite. The persistent zone of semischist adjacent to the amphibolite is interpreted to be a zone of cataclasis and low-grade metamorphism at the base of the plate of sedimentary rocks.

THE MALLETHEAD THRUST

The Duzel Phyllite and Sissel Gulch Graywacke override the unmetamorphosed Moffett Creek Formation on the Mallethead thrust, named by Churkin and Langenheim (1960) for Mallethead Rock, a prominent topographic feature in the China Mountain quadrangle. Presumably the Duzel and Sissel Gulch Formations also overrode the Gazelle Formation in the Lime Gulch area but have since been removed by erosion.

The Mallethead thrust is a major structural feature of the eastern Klamath plate. In the Lime Gulch area its sinuous trace indicates that it dips gently westward and, being slightly warped, reappears on the west side of Scarface Ridge. Churkin and Langenheim (1960) interpreted it to have a broad north-plunging anticlinal form in the Lime Gulch area that was breached by erosion. The relatively straight trace on the east, however, indicates a steeply inclined to vertical fault, which is interpreted here as a normal fault that intersects the Mallethead thrust and drops it down on the east. The thrust window that exposes greenschist and other metamorphic rocks and trondhjemitic bodies beneath the Sissel Gulch Graywacke and Duzel Phyllite south of Sissel Gulch is in the Mallethead thrust. Mapping by myself and others (D. M. Rohr, written commun., 1974; Zdanowicz, 1971) shows that the Mallethead thrust can be traced to the southwest where it overrides the same graywacke and phyllite in the China Mountain and Etna quadrangles.

OTHER THRUST FAULTS

The contact between the Duzel Phyllite and the Schulmeyer Gulch sequence where it crosses the ridge between Moffett Creek and Cottonwood Creek is interpreted as a thrust fault, with the Duzel on the upper plate. The contact, which is well defined by an abrupt change in lithology, is accompanied by brecciation and much shearing and some local silicification. At least two small bodies of quartz-bearing diabase or gabbro occur on the contact.

The Antelope Mountain Quartzite overlies the Schulmeyer Gulch sequence with a gently eastward-dipping contact which is interpreted as a fault. The upper plate of Antelope Mountain Quartzite overrides the fault between the Duzel Phyllite and the Schulmeyer Gulch sequence and has the form of a broad, shallow syncline that plunges gently northward. Probably the Antelope Mountain Quartzite has not been displaced a long distance, for its eastern contact with the Duzel is depositional, modified by high-angle faulting in places.

The belt of Moffett Creek Formation in the western part of the Yreka quadrangle occupies a thrust plate that extends from near Yreka almost to Callahan in the Etna quadrangle. The unmetamorphosed but highly disturbed calcareous siltstone unit is thrust over the Duzel Phyllite and the Schulmeyer Gulch sequence. The plate is apparently synclinal with a steep western contact and a gentler west-dipping eastern contact. Its position over the Duzel is puzzling, for the formation is overridden by the Duzel Phyllite and the Sissel Gulch Graywacke on the Mallethead thrust plate.

The uppermost thrust plate is represented by the small limestone klippe overlying the Moffett Creek Formation. A narrow gouge zone is exposed at the base of the limestone cliff on the east side of Duzel Rock, and limestone beds and interbedded greenstone strata are cut off at the nearly flat, slightly curved contact. Similar isolated limestone bodies in the Etna quadrangle mapped by Romey (1962) are probably klippen that are remnants of a formerly more extensive thrust plate.

FOLDING

A foliation has been developed in rocks of the eastern Klamath belt that are above the Mallethead thrust and beneath the thrust plate of Moffett Creek Formation. The foliation surface, which is parallel to the original bedding, is developed best in the Duzel Phyllite, whose original rocks were mostly pelitic; it is also commonly seen in the Schulmeyer Gulch sequence and in some facies of the Antelope Mountain Quartzite; it is developed less strongly in most exposures of the Sissel Gulch Graywacke. The foliation, manifested by the alinement of micaceous minerals parallel to the bedding, was developed during an episode of low-grade regional metamorphism. In a few places the foliation is apparently an axial plane cleavage, but the general orientation of the folding direction is not known.

A second episode of folding is expressed, mainly in the Duzel Phyllite and the Schulmeyer Gulch sequence, by the superposition on the foliation of closely spaced small-amplitude folds that are apparent in nearly every exposure and give the rocks a character-istic crumpled appearance. The axial planes of these folds have predominantly northerly strikes and mainly eastward dips, much less commonly to the west. The axes plunge gently northward or southward, and many are subhorizontal, but a few have a moder-ate or steep plunge.

Broad open folds involving entire thrust plates in the eastern Klamath belt are representative of a third folding episode. Thus, the Antelope Mountain Quartzite is in a syncline, where as anti-clinal folds in the southern part of the plate bearing the Duzel Phyllite and Sissel Gulch Graywacke expose the Gazelle Forma-tion in the Lime Gulch area and low-grade schist and trondhjemitic rocks along Moffett Creek. The plate of Moffett Creek Formation is synclinal in its northern part, and in the southwest part of the Yreka quadrangle it is warped into an anticline flanked by two synclines.

STRUCTURE OF THE GAZELLE FORMATION

The Gazelle Formation is not deformed and metamorphosed like the Duzel Phyllite and Sissel Gulch Graywacke. A broad anticline can be recognized in the Lime Gulch area, and southwestward in the China Mountain quadrangle an open, southwest-plunging syncline is mapped. Widespread faulting, more evident in the China Mountain than in the Yreka quadrangle, has disrupted the lower member of the Gazelle and in places has apparently resulted in severe dismemberment of the unit, to the extent that the Gazelle may be considered a broken formation.

STUART FORK THRUST PLATE

The Stuart Fork Formation is thrust northward and northwestward over the Permian greenstone-chert assemblage on the Soap Creek Ridge thrust fault in the northwestern part of the Yreka quadrangle. The Soap Creek Ridge thrust emerges from beneath the Hornbrook Formation and Tertiary volcanic rocks in the Hornbrook quadrangle and extends southwestward into the Yreka quadrangle where it intersects the Greenhorn fault north of Yreka (Masson, 1949). The thrust fault resumes near Greenhorn Creek after an interruption of 7 km (4.3 mi) and continues southwestward about 40 km (25 mi). The dip is to the southeast at low to moderate angles.

The thrust plate is limited on the southeast by the northeast-trending serpentinite body. The contact apparently is a fault whose dip ranges from vertical to steeply eastward.

The structure within the Stuart Fork plate has not been investigated in detail, but exposures of the predominant phyllitic quartzite commonly show an intricate small-scale folding that gives the rocks a crumpled appearance. A shear-surface foliation apparently parallel to the axial plane of isoclinal or nearly isoclinal folds strikes predominantly north and dips gently to moderately eastward. The foliation is usually parallel to the compositional layering or bedding. Lineations in this surface plunge northeast or southwest at low to moderate angles. A second direction of folding is commonly visible and in many places is responsible for the crumpled appearance of outcrops of phyllitic quartzites in the Stuart Fork Formation. This folding episode is represented by small-amplitude folds that cross the earlier foliation at varying angles, but commonly the axial planes of the folds also have a northerly strike and dip steeply eastward. The axes plunge northward or southward at small to moderate angles. The folds range from open, simple corrugations of the foliation to complex, commonly tightly appressed asymmetric wrinkles. The scattered tectonic blocks of metabasalt in the Stuart Fork might be interpreted as indicative of a melange. The enclosing metasedimentary rocks have also undergone high P/T metamorphism, however, so if tectonic mixing took place, it was prior to the episode of blueschist metamorphism. The basaltic rocks may have been part of the normal succession and were disrupted during the metamorphism.

TECTONIC HISTORY

The earliest tectonic episode recorded in the rocks of this region is the uplift of an Early Ordovician ophiolite sequence of upper mantle and oceanic crust to provide detritus to Silurian sediments represented here by the Gazelle Formation. Burchfiel and Davis (1975) suggested that such an uplift may have been a geanticlinal buckling of the sea floor adjacent to an incipient trench. Elsewhere, possibly eastward, at about the same time or possibly earlier, another area was uplifted, probably a volcanic arc, which provided andesitic volcanic debris to the Sissel Gulch Graywacke. Sediments from a sialic continental source or a more deeply eroded arc were incorporated in the Duzel Phyllite, Antelope Mountain Quartzite, and Moffett Creek Formation.

The time and place of disruption of the Moffett Creek Formation are obscure. A. W. Potter (written commun., 1975) reported seeing a block of typical Moffett Creek Formation—clasts of siltstone dispersed in a sheared shale matrix—enclosed in a bouldery-pebbly mudstone debris flow unit of the Gazelle Formation, which is evidence that disruption of the Moffett Creek took place during or before Silurian time. The time of deformation and low-grade regional metamorphism of the Duzel Phyllite and Sissel Gulch Graywacke are also uncertain. D. M. Rohr (written commun., 1975) found clasts of phyllite similar to those of the Duzel Phyllite in Early Devonian rocks in the China Mountain quadrangle, suggesting that metamorphism of the Duzel and Sissel Gulch may have occurred during disruption of the Moffett Creek at a different place. The potassium-argon date obtained on semischist adjacent to the amphibolite is supportive evidence for Early Silurian tectonism.

A well-defined Middle Devonian tectonic episode was responsible for metamorphism of the Salmon and Abrams (or Grouse Ridge) Formations in the southern Klamath Mountains, probably represented by the amphibolite belt in the Yreka quadrangle and amphibolite and associated schistose marble in the Fort Jones and Etna quadrangles. This is compatible with a postulated Middle Devonian orogenic interval of regional extent in the western Cordillera of North America (Boucot and others, 1974).

Tectonism in Early and Middle Triassic time is recorded in the Stuart Fork Formation, which was metamorphosed in a tempera-.ture-pressure environment favorable for the development of the blueschist metamorphic facies. The Stuart Fork is between the eastern Klamath plate of lower Paleozoic sedimentary rocks and the greenstone-chert assemblage of Permian age and thus may have been metamorphosed by the collision of an oceanic plate and island-arc or continental rocks. Blueschists of similar age occur in eastern Oregon (M. A. Lanphere, written commun., 1974) and in the Pinchi Lake area, central British Columbia (Paterson and Harakal, 1974). Evidence of Early or Middle Triassic tectonism in the Sierra Nevada is found in the folding of Permian and older sedimentary and volcanic rocks in the Ritter Range pendant directly west of granitic rocks emplaced during the Lee Vining intrusive epoch 210 m.y. ago (Evernden and Kistler, 1970). Mattinson (1972) has shown that quartz diorites about 220 m.y. old are present in the Marblemount belt in the northern Cascade Range. These data suggest a widespread early Mesozoic orogenic episode in the western Cordillera, perhaps related to the major later Mesozoic orogeny.

The age of thrusting of the Lower Paleozoic sedimentary rocks cannot be clearly established. Some thrusting could have been Early Devonian, but later episodes of faulting could also have occurred (for example, during Early and Middle Triassic, and perhaps as late as the widespread Jurassic Nevadan orogeny). In the Yreka quadrangle the gabbroic to diabasic dikes that intrude the Paleozoic rocks are possible manifestations of this orogeny. No granitic plutons of Jurassic age occur here, but they are common in adjoining terranes. The Late Cretaceous marine strata of the Hornbrook Formation that were deposited unconformably on thrust plates of the older rocks show that thrusting occurred before this time.

Evidence of Cenozoic diastrophism is found in the remnants of uplifted volcanic rocks in the southern and eastern parts of the quadrangle.

MINERAL DEPOSITS

Gold has been the chief metallic commodity of economic interest in the Yreka quadrangle, and a small amount of chrome ore has also been produced. The principal nonmetallic commodity is limestone, which has been actively mined on a small scale for several years.

GOLD

Gold production has been restricted to the small area in the northwest corner of the Yreka quadrangle occupied by rocks of the western Paleozoic and Triassic belt, bounded on the south by the Greenhorn fault and the Soap Creek Ridge thrust fault. No gold is known to occur in the eastern Klamath and central metamorphic belts. All the properties, both lodes and placers, are now idle.

Gold was discovered on the flats west of Yreka in March 1851, and soon the surrounding area was being actively prospected and mined. Most of the production came from placers on Greenhorn Creek, the alluvial flats west of Yreka Creek below its confluence with Greenhorn Creek, and from Yreka Creek itself. Lesser quantities of silver were also recovered. In 1933 dredging operations commenced on Greenhorn Creek and continued until 1939, successfully recovering gold from ground that had previously been worked by hand. Other dredge operations were conducted on Yreka Creek until late 1942 when all gold mining operations ceased with the issuance of War Production Board Gold Mine Closing Order L-208. More than 10,000 oz of placer gold and more than 1,000 oz of silver were recovered in the decade 1933-43 (U.S. Bureau Mines, 1934-43). Probably little remains to be won from the alluvium in the area, certainly not economically.

The Hornbrook Formation has an unknown potential as a source of placer gold. The Blue Gravel mine near Henley, about 19 km (12 mi) north of Yreka, yielded some gold from a weathered basal conglomerate of the Hornbrook (Dunn, 1894). The Hornbrook Formation in the Yreka quadrangle may also contain placer gold, but its value would be difficult to determine, and recovery would be difficult and expensive.

Prospecting on the slopes adjoining the creeks resulted in the discovery of lode deposits on which several prospects and small mines were established. The gold occurred free in quartz veins that filled fractures in the metavolcanic rocks. Pyrite commonly accompanied the gold, and galena and sphalerite were also present in small amounts locally.

The Osgood (Lucky Strike) mine discovered in 1909-10 about 3 km (2 mi) west of Yreka in NW¼ sec. 21, T. 45 N., R. 7 W., was variously reported to have yielded more than $10,000 to more than $40,000 in gold from several north-south vertical veins (Averill, 1931; Brown, 1916; Logan, 1925).

The Mount Vernon mine at the head of Greenhorn Creek is in fragmental greenstone on the south side of the Greenhorn fault, in NW¼ sec. 26, T. 45 N., R. 8 W. Its date of discovery is unknown, but it was worked intermittently for more than 50 years and was in operation as late as 1937. Production is unknown, but several hundred feet of workings on at least five levels were developed on a northwest-striking east-dipping vein system of ribbon quartz as much as 0.6 m (2 ft) wide (Brown, 1916; Logan, 1925; Averill, 1931, 1935; O'Brien, 1947).

The Katie May, a small mine originally located in 1886 somewhere near the Scorpion mine (pl. 1) in sec. 14, T. 45 N., R. 8 W., reportedly yielded more than $70,000 in gold (Averill, 1935) from narrow quartz veins and stringers in greenstone near a black shale contact. No production data are available for the Scorpion mine.

CHROMITE

Chromite, the ore of chromium, has been mined from deposits in serpentinite in the western part of the Yreka quadrangle (Wells and Cater, 1950). The deposits, which were podlike bodies of massive chromitite in serpentinite—usually serpentinized dunite—occurred north of Moffett Creek in NE¼ sec. 26, T. 44 N., R. 8 W. near the Littrel mine (plate 1), on the south side of the creek at the Pegleg mine (SE¼ sec. 26, T. 44 N., R. 8 W.), and a cluster of pods called the Moffett Creek group occurred in the southern part of sec. 35, T. 44 N., R. 8 W. The main production from these deposits was in 1942–43, when practically all the visible pods of massive ore were mined out so that it was impossible to make any estimation of reserves (Wells and Cater, 1950).

LIMESTONE

Limestone has been quarried sporadically for an unknown period from the Payton Ranch Limestone Member of the Gazelle Formation west of Gazelle in Lime Gulch. Several times during the present investigation quarrying was in progress here.

Large reserves of limestone are present in an easily accessible area in Lime Gulch. The limestone is of good quality, high in $CaCO_3$ and low in MgO as shown by analyses reported by Heyl and Walker (1949). The principal use has been for agricultural limestone, although it has also been shipped for treatment as carbide rock and for the manufacture of paper.

Numerous bodies of limestone also occur in the Schulmeyer Gulch sequence, and some of the bodies are fairly accessible. One body in NE¼ sec. 4, T. 44 N., R. 7 W. has been quarried for road metal. The purity of the limestone from these bodies is not known because they have not been sampled.

REFERENCES CITED

Averill, C. V., 1931, Preliminary report on economic geology of the Shasta quadrangle: California Jour. Mines and Geology, v. 27, no. 1, p. 3-65.
———1935, Mines and mineral resources of Siskiyou County: California Jour. Mines and Geology, v. 31, no. 3, p. 255-238.
Bailey, E. H., Irwin, W. P., and Jones, D. L., 1964, Franciscan and related rocks, and their significance in the geology of western California: California Div. Mines and Geology Bull. 183, 177 p.
Boucot, A. J., Dunkle, D. H., Potter, A., Savage, N. M., and Rohr, D., 1974, Middle Devonian orogeny in western North America? A fish and other fossils: Jour. Geology, v. 82, p. 691-708.
Brown, G. C., 1916, Mines and mineral resources of Shasta, Siskiyou, and Trinity Counties, California: California Mining Bur., Chapters of State Mineralogists Rept. 1913-14, p. 745-924.
Burchfiel, B. C., and Davis, G. A., 1975, Nature and controls of Cordilleran orogenies, Western United States: Extensions of an earlier synthesis: Am. Jour. Sci., v. 275-A, p. 363-396.
Churkin, Michael, Jr., 1965, First occurrence of graptolites in the Klamath Mountains, California, in Geological Survey research 1965: U.S. Geol. Survey Prof. Paper 525-C, p. C72-C73.
Churkin, Michael, Jr., and Langenheim, R. L., Jr., 1960, Silurian strata of the Klamath Mountains, California: Am. Jour. Sci., v. 258, no. 4, p. 258-273.
Coleman, R. G., and Lee, D. E., 1963, Glaucophane-bearing metamorphic rock types of the Cazadero area, California: Jour. Petrology, v. 4, no. 2, p. 260-301.
Condie, K. C., and Snansieng, Sathian, 1971, Petrology and geochemistry of the Duzel (Ordovician) and Gazelle (Silurian) Formations, northern California: Jour. Sed. Petrology, v. 41, no. 3, p. 741-751.
Davis, G. A., 1968, Westward thrust faulting in the south-central Klamath Mountains, California: Geol. Soc. America Bull., v. 79, no. 7, p. 911-934.
Davis, G. A., Holdaway, M. J., Lipman, P. W., and Romey, W. D., 1965, Structure, metamorphism, and plutonism in the south-central Klamath Mountains, California: Geol. Soc. America Bull., v. 76, no. 8, p. 933-965.
Davis, G. A., and Lipman, P. W., 1962, Revised structural sequence of pre-Cretaceous metamorphic rocks in the southern Klamath Mountains, California: Geol. Soc. America Bull., v. 73, no. 12, p. 1547-1552.
Diller, J. S., 1886, Notes on the geology of northern California: U.S. Geol. Survey Bull. 33, 23 p.
Diller, J. S., and Schuchert, Charles, 1894, Discovery of Devonian rocks in California: Am. Jour. Sci., v. 47, ser. 3, p. 416-422.
Dunn, R. L., 1894, Auriferous conglomerate in California: California Div. Mines, Rept. 12, p. 459-471.
Elliott, M. A., 1971, Stratigraphy and petrology of the late Cretaceous rocks near Hilt and Hornbrook, Siskiyou County, California, and Jackson County, Oregon [abs.]: Dissert. Abs. Internat., v. 32, no. 2, p. 1021B.
Elliott, M. A., and Bostwick, D. A., 1973, Occurrence of *Yabeina* in the Klamath Mountains, Siskiyou County, California: Geol. Soc. America Abs. with Prog., v. 5, no. 1, p. 38.

Evernden, J. F., and Kistler, R. W., 1970, Chronology of emplacement of Mesozoic batholithic complexes in California and western Nevada: U.S. Geol. Survey Prof. Paper 623, 42 p.

Fyfe, W. S., and Turner, F. J., 1966, Reappraisal of the metamorphic facies concept: Contr. Mineralogy and Petrology, v. 12, p. 354-364.

Hershey, O. H., 1901, Metamorphic formations of northwestern California: Am. Geologist, v. 27, p. 225-245.

Heyl, G. R., and Walker, G. W., 1949, Geology of limestone near Gazelle, Siskiyou County, California: California Jour. Mines and Geology, v. 45, no. 4, p. 514-520.

Hopson, C. A., and Mattinson, J. M., 1973, Ordovician and Late Jurassic assemblages in the Pacific northwest: Geol. Soc. America Abs. with Prog., v. 5, no. 1, p. 57.

Hotz, P. E., 1967, Geologic map of the Condrey Mountain quadrangle, and parts of the Seiad Valley and Hornbrook quadrangles, California: U.S. Geol. Survey Geol. Quad. Map GQ-618, scale 1:62,500.

———1973, Blueschist metamorphism in the Yreka-Fort Jones area, Klamath Mountains, California: U.S. Geol. Survey Jour. Research, v. 1, no. 1, p. 53-61.

Hotz, P. E., Lanphere, M. A., and Swanson, D. A., 1977, Blueschist of Triassic age from northern California and north-central Oregon: Geol. Soc. America, Geology (in press).

Hsü, K. J., 1968, Principles of mélanges and their bearing on the Franciscan-Knoxville paradox: Geol. Soc. America Bull., v. 79, no. 8, p. 1063-1074.

Irwin, W. P., 1960, Geologic reconnaissance of the northern Coast Ranges and Klamath Mountains, California, with a summary of the mineral resources: California Div. Mines Bull. 179, 80 p.

———1966, Geology of the Klamath Mountains province, in Bailey, E. H., ed., Geology of northern California: California Div. Mines and Geology Bull. 190, p. 19-38.

———1972, Terranes of the western Paleozoic and Triassic belt in the southern Klamath Mountains, California, in Geological Survey research 1972: U. S. Geol. Survey Prof. Paper 800-C, p. C103-C111.

———1973, Sequential minimum ages of oceanic crust in accreted tectonic plates of northern California and southern Oregon: Geol. Soc. America Abs. with Prog., v. 5, no. 1, p. 62-63.

Irwin, W. P., and Lipman, P. W., 1962, A regional ultramafic sheet in eastern Klamath Mountains, California, in Geological Survey research 1962: U. S. Geol. Survey Prof. Paper 450-C, p. C18-C21.

Jones, D. L., 1959, Stratigraphy of Upper Cretaceous rocks in the Yreka-Hornbrook area, northern California [abs.]: Geol. Soc. America Bull., v. 70, no. 12, pt. 2, p. 1726-1727.

Kuno, Hisashi, 1968, Differentiation of basalt magmas, in Hess, H. H., and Poldervaart, A., eds., Basalts—The Poldervaart treatise on rocks of basaltic composition, V. 2: New York, Interscience Publishers, p. 623-688.

LaFehr, T. R., 1966, Gravity in the eastern Klamath Mountains, California: Geol. Soc. America Bull., v. 77, no. 11, p. 1177-1189.

Lanphere, M. A., Irwin, W. P., and Hotz, P. E., 1968, Isotopic age of the Nevadan orogeny and older plutonic and metamorphic events in the Klamath Mountains, California: Geol. Soc. America Bull., v. 78, no. 8, p. 1027-1052.

Lindsley-Griffin, Nancy, 1973, Lower Paleozoic ophiolite of the Scott Mountains, eastern Klamath Mountains, California: Geol. Soc. America Abs. with Prog., v. 5, no. 1, p. 71-72.

Logan, C. A., 1925, Geology and mineral resources of Siskiyou County: California

Jour. Mines and Geology, v. 21, no. 4, p. 413–498.

Mack, Seymour, 1958, Geology and groundwater features of Scott Valley, Siskiyou County, California: U.S. Geol. Survey Water-Supply Paper 1462, 98 p.

——1960, Geology and groundwater features of Shasta Valley, Siskiyou County, Calif.: U.S. Geol. Survey Water-Supply Paper 1484, 115 p.

Masson, P. H., 1949, Geology of the Gunsight Peak district, Siskiyou County, California: California Univ., Berkeley, M.A. thesis, 74 p.

Mattinson, J. M., 1972, Ages of zircons from the northern Cascade Mountains, Washington: Geol. Soc. America Bull., v. 83, p. 3769–3784.

Merriam, C. W., 1940, Devonian stratigraphy and paleontology of the Roberts Mountains region, Nevada: Geol. Soc. America Spec. Paper 25, 114 p.

——1972, Silurian rugose corals of the Klamath Mountains region, California: U.S. Geol. Survey Prof. Paper 738, 50 p.

Moore, J. G., 1965, Petrology of deep-sea basalt near Hawaii: Am. Jour. Sci., v. 263, no. 1, p. 40–52.

O'Brien, J. C., 1947, Mines and mineral resources of Siskiyou County: Calif. Jour. Mines and Geology, v. 43, no. 4, p. 413–462.

Patterson, I. A., and Harakal, J. E., 1974, Potassium-argon dating of blueschists from Pinchi Lake, central British Columbia: Canadian Jour. Earth Sci., no. 11, p. 1007–1011.

Peck, D. L., Imlay, R. W., and Popenoe, W. P., 1956, Upper Cretaceous rocks of parts of southwestern Oregon and northern California: Am. Assoc. Petroleum Geologists Bull., v. 40, p. 1968–1984.

Porter, R. W., 1973, Geology of the Facey Rock area, Etna quadrangle, California: Oregon State Univ., M.A. thesis, 87 p.

Potter, A. W., and Boucot, A. J., 1971, Ashgillian, Late Ordovician brachiopods from the eastern Klamath Mountains of northern California: Geol. Soc. America Abs. with Progs., v. 3, no. 2, p. 180.

Potter, A. W., Hotz, P. E., and Rohr, D. M., 1977, Stratigraphy and inferred tectonic framework of Lower Paleozoic rocks in the eastern Klamath Mountains, northern California, in Paleozoic paleogeography of the western United States: Soc. Econ. Paleontologists and Mineralogists, Pacific Sec., Pacific Coast Paleogeography Symposium 1, p. 421–440.

Rohr, David, and Boucot, A. J., 1971, Northern California (Klamath Mountains) pre-Late Silurian igneous complex: Geol. Soc. America Abs. with Progs., v. 3, no. 2, p. 186.

Romey, W. D., 1962, Geology of a part of the Etna quadrangle, Siskiyou County, California: California Univ., Berkeley, unpub. Ph. D. thesis, 93 p.

Shapiro, Leonard, 1967, Rapid analysis of rocks and minerals by a single solution method, in Geological Survey research 1967: U.S. Geol. Survey Prof. Paper 575-B, p. B187–B191.

U.S. Bureau of Mines, 1934–43 [issued annually], Minerals yearbook, 1934–43: U.S. Dept. Interior, Bur. Mines.

Wells, F. G., and Cater, F. W., Jr., 1950, Chromite deposits of Siskiyou County, Calif., in Geological investigations of chromite in California: Calif. Div. Mines and Geology, Bull. 134, pt. I, ch. 2, p. 77–127.

Wells, F. G., Walker, G. W., and Merriam, C. W., 1959, Upper Ordovician(?) and Upper Silurian formations of the northern Klamath Mountains, California: Geol. Soc. America Bull., v. 70, no. 5, p. 645–649.

Williams, Howel, 1949, Geology of the Macdoel quadrangle [Calif.]: California Div. Mines and Geology Bull., v. 151, p. 7–60.

72   REFERENCES CITED

Williams, Howel, Turner, F. J., and Gilbert, C. M., 1955, Petrography—An introduction to the study of rocks in thin sections: San Francisco, Calif., W. H. Freeman and Co., 406 p.

Yoder, H. S., Jr., 1967, Spilites and serpentinites: Carnegie Inst. Wash. [D.C.] Yearbook 65, 1965–66, p. 269–279.

Zdanowicz, Ted, 1971, The folded Mallethead thrust, eastern Klamath Mountains, California: Geol. Soc. America Abs. with Prog., v. 3, no. 2, p. 223.

www.ingramcontent.com/pod-product-compliance
Lightning Source LLC
Chambersburg PA
CBHW081559170526
45166CB00009B/2756